PATRICK CAMPBELL left his home and family run hotel in Donegal to work as a tunnel tiger in Scotland before emigrating to America. He was marketing communications manager for the World Trade Center until he retired in 1996, after a marketing career with the Port Authority of New York and New Jersey that lasted 35 years. He is a graduate of Rutgers University with a Masters Degree in English and he owns an internet bookshop, P.H. Campbell, which specialises in first editions of books on Irish history, literature and folklore.

He was formerly a theatre critic, book editor and columnist and has appeared in documentaries and numerous television and radio shows. Campbell was born in Dungloe, Co Donegal. He lives in Jersey City, New Jersey, with his Donegal-born wife. He has one daughter.

By the same author:

A Molly Maguire Story

Memories of Dungloe

Death in Templecrone

The Last Days of Oscar Devenney

Who Killed Franklin Gowen?

Tunnel Tigers

A First Hand Account of a Hydro Boy in the Highlands

PATRICK CAMPBELL

Luath Press Limited

EDINBURGH

www.luath.co.uk

ACKNOWLEDGEMENTS

The following people have provided invaluable input into this book:

Sean and Breige Boyle, Donegal, Ireland.
Jimmie Campbell, Donegal, Ireland.

Michael Campbell, New Jersey.
Hugh O'Donnell, New Jersey.
Eileen and Nora Campbell, New Jersey.

Eric McKeever, Maryland.

Christine Goldbeck, Pennsylvania.
Howard Crown, Pennsylvania.

Alastair Carmichael, Jr.

First published 2000, by P. H. Campbell, USA
New edition 2005
Reprinted 2008

ISBN (10): 1-842820-72-9
ISBN (13): 978-1-842820-72-8

The paper used in this book is recyclable. It is made from low chlorine pulps produced in a low energy, low emission manner from renewable forests.

Printed and bound by Bell & Bain Ltd., Glasgow

Typeset in 10.5pt Sabon by 3btype.com

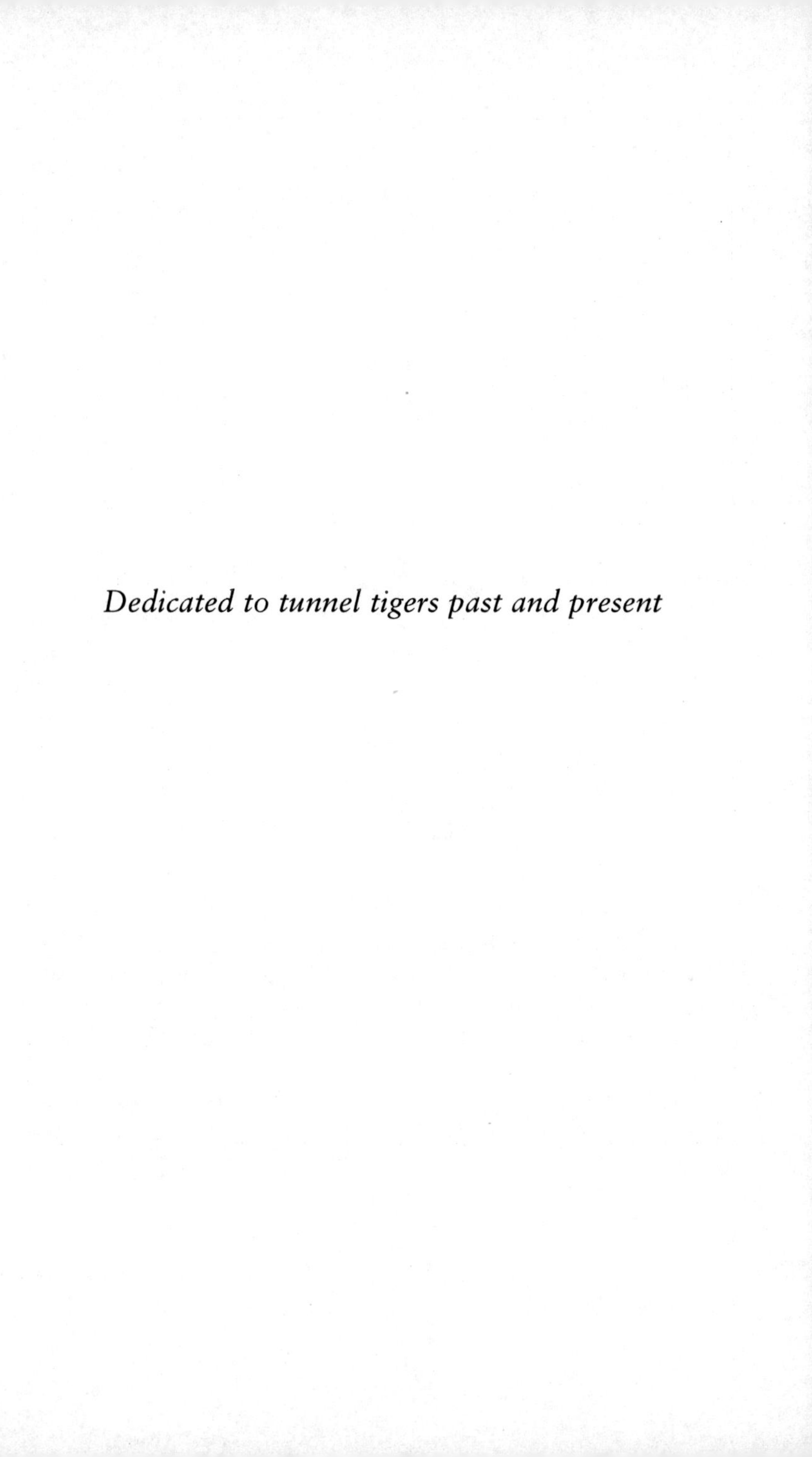

Dedicated to tunnel tigers past and present

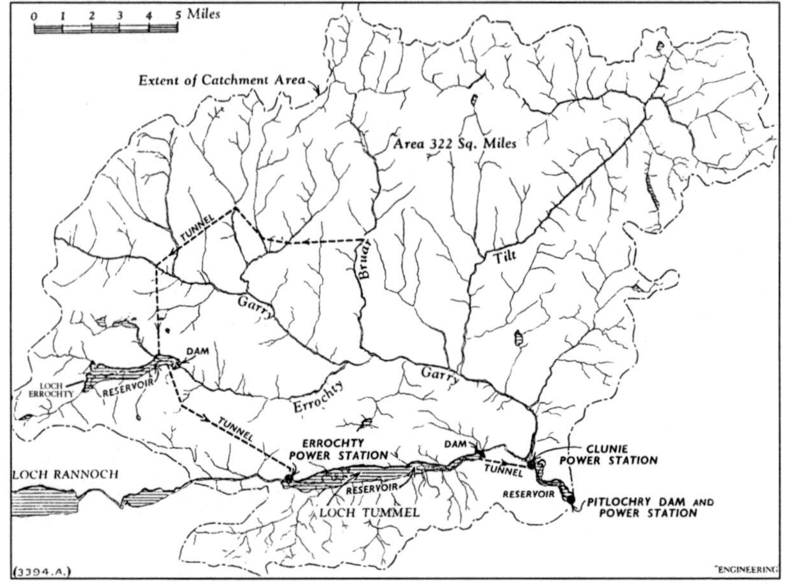

A note from the author

Tunnel Tigers is an account of the author's experience working in the Highlands of Scotland in the early 1950s.

Other Irish workers may have had a different experience on different projects, and the anti-Irish prejudice noted by the author may have varied in intensity in other parts of Scotland during this period.

Some of the personalities described here have been given fictitious names, but the incidents described are accurate.

Contents

Foreword

THE REMOTE WILDERNESSES, mountains and glens of the Scottish Highlands had not experienced such an invasion of outsiders since the government troops had arrived to subdue and control the rebellious Jacobite clans in the early 1700s. This next intrusion in the 1940s and 1950s was somewhat less hostile, although it had followed an ultimately futile attempt by landowners and others to prevent significant changes in Scotland's landscape. The Hydro Men had arrived in their thousands from other parts of Scotland, Ireland, England, Canada and Australia, accompanied by no small number of German POWs, Poles, Czechs, and Ukrainians, to name a few. Their task was to construct several hydro electric schemes, thereby harnessing Scotland's plentiful water supply and producing economic, clean power for not only the local regions but also the south of Scotland and north of England.

One of the most prominent civil engineering contractors involved in these huge projects for the North of Scotland Hydro Electric Board was A. M. Carmichael Ltd of Edinburgh, founded by my grandfather around 1914.

The Highlands' rugged landscape and extremely harsh weather proved a grueling test for the men charged with bringing these projects to completion, which involved hundreds of miles of tunneling through mountains, construction of massive dams creating new lochs and building aqueducts, roads and power stations.

One Donegal man, who worked for Carmichael on the Errochty section of the Tummel-Garry scheme, took it upon himself to write a most fascinating and detailed account of his reasons for the move from Ireland to the lucrative work on the hydro schemes. His experiences both on and off site and portrayal of the many interesting and colourful characters he worked with is captivating.

I commend Patrick Campbell most highly on his book *Tunnel Tigers*, which provides such a splendid insight into the reality of the men and the situations experienced by thousands but documented by only one.

At Errochty, Carmichael employed some 2,000 men from 1948 to 1957 and the work involved over 18 miles of tunnel as well as aqueducts, roads, an underground power station, and the second largest buttress dam in Scotland – 1,386 feet long and 120 feet high. The largest, at Glen Shira in Argyll, was also being built by A. M. Carmichael at the same time.

Alastair M. Carmichael
(Son of Alastair M. Carmichael (director of A. M. Carmichael Ltd) and grandson of Alastair M. Carmichael (founder of A. M. Carmichael Ltd)).

Introduction

VERY FEW PEOPLE WHO grew up in Dungloe, County Donegal in the late 1940s and early 1950s ever planned to continue living there once they grew old enough to be considered an adult.

There was no reason for them to consider this unless they were the son or daughter of a businessman and were in line to inherit the family business. Jobs were few and far between during this period, and the only income available was a meagre dole which would barely ward off starvation for those who relied on it.

So, unless they were lucky enough to have parents who could afford to send them away to high school in nearby Letterkenny, they began to plan their getaway once they finished primary school at the age of fourteen.

Abandoning the nest at the age of fourteen or fifteen was considered normal in those days. Indeed, the parents who were raising children in the Forties would tell their brood that when they themselves were young they were shipped off at the age of ten to East Donegal or County Tyrone to work for farmers for six months at a time, and that, by the time they were fifteen, they were considered young adults, not children.

Dungloe and the surrounding mountains and islands have been an incubator for a very long time, for centuries, in fact. In this area large families were raised to provide labour for Britain, the United States, Canada and Australia. Most of this labour was unskilled, and most of the jobs filled abroad were the jobs that no one else wanted, jobs like household help, digging potatoes, or digging ditches.

Only one or two members of each family remained at home to take care of their parents and grandparents. Those who stayed at home raised large families of their own and continued the cycle, or they became the classic Irish bachelor or spinster of Irish folklore.

My own ancestors were heavily involved in emigration as far back as the beginning of the 19th century. My great grandfather, Donal Sweeney, and his brother, John, emigrated to New York City in the 1830s, where they found work in construction.

Donal did not like living in America, however, and he returned to Ireland after five years. John however had heard that there were fortunes being made out West in the fur trade with the Native Americans and he headed out to the Missouri River, where he spent ten years living and trading with the Indians and accumulating a small fortune, which he invested in Sweeney's Hotel, Dungloe, in the early 1840s. The hotel became one of the best known establishments in northwest Donegal.

My father's uncle, Alec Campbell, emigrated to Pennsylvania in 1868, accompanied by his brothers, Teague and James, and his sisters, Annie and Sarah. The Campbell siblings moved into a home in Tamaqua occupied by their uncle, who was also named Alec. The older Alec had fled Ireland during the famine of the 1840s and had established himself as a coal miner.

The three Campbell brothers also went to work in the coal mines, but their story was not to be the success story of John Sweeney trading fur with the Indians – the Campbells became involved in a deadly confrontation with railroad and coal mining baron Franklin Gowen, and Alec wound up on the gallows, executed for being an accessory to the murder of coal mine boss Jack Jones. Alec maintained his innocence to the end; James and Teague fled to Australia.

Tim Gallagher, my grandfather on my mother's side of the family, emigrated for a brief period to Scotland, but eventually he came home and married Brigid Doherty from Dungloe, and both lived in Innisfree for the next seventy-five years, raising seven children, all of whom eventually emigrated to America. He supported his family by fishing, and growing vegetables that he sold on the mainland.

Tim and Brigid were raised in Gaelic-speaking households, but neither of them taught Gaelic to their children, who spoke only English. It was Tim's opinion that all his children would have to go abroad to earn a living and that Gaelic would be of no use to them on the streets of New York or London. So, he made sure they were fluent in English.

Tim and Brigid both came from large families, but the vast majority of their siblings emigrated to the American Midwest, and their descendants can be found today scattered all over Michigan, Idaho, Minnesota and Illinois.

My father followed his ancestors to America in the 1890s after having worked for several years in Scotland. He initially went to Pennsylvania and worked in the coal mines, but after several years he found a job working for Standard Oil in Elizabeth, New Jersey.

In 1914, when he was forty-three years old, he returned to Ireland on holiday and fell in love with my mother, who was 21 years old. They married and he brought her back to Elizabeth, NJ, where the three oldest children were born.

But my mother pined for Ireland, and eventually she persuaded my father to invest his savings in the purchase of an hotel on Main Street, Dungloe, where another six children were born. I was the youngest, born there in 1934.

Even though my parents were quite content to live out their lives in Dungloe, one by one their offspring headed out for greener pastures. Brigid Mary, Annie and Catherine married and went off to live in the southern part of Ireland; Rose and Theresa went off to Scotland; Bernie joined the American Merchant Marine and voyaged all over the world; and James joined the American Army in the closing months of World War II and marched into defeated Germany with the American troops.

By 1951, I was the only sibling left, and I also had a strong urge to leave the nest and travel to faraway places.

The stories told by Bernie and James about their adventures abroad had a great deal to do with my urge to cut my ties with Dungloe. James told me fascinating stories about his experiences in wartime Germany, and Bernie had a whole array of tales about being a radio operator on American merchant ships that were a prime target for German submarines.

I do not know how many of Bernie's stories were based on fact and how many were the product of an overripe imagination, but I soaked them all up anyway and was jealous of the adventurous life he had led, all before he had reached the age of 21.

But my brothers' stories were only part of the reason for my growing hunger for adventures abroad. All my life, up until that point, my father had told me stories about the adventures of my great-great-uncle, John Sweeney, among the American Indians, and about the execution of his uncle in Pennsylvania.

My father was a gifted storyteller and he could mesmerise an audience in our bar as he told his customers about great-uncle John hunting buffalo on the plains of Kansas and trapping other wild animals deep in the wilderness. The American West began at St Louis in the early 19th Century and Indian villages existed side by side with white settlements in Kansas. Once John crossed the Missouri River at St Joseph, Kansas, however, he was in Indian country, and he remained in Indian country all the way to Oregon and California.

Inspired by my father's stories about John Sweeney's adventures in the wilderness, I went to the local library and read every book I could about the conquest of the American West. I learned that Sweeney's years in the West coincided with the beginning of the end for the Native Americans. First Missouri was taken, then Kansas, then all the buffalo were slaughtered, and finally in the 1870s and 1880s, the Indians were herded into reservations and the 'West was won' for the white man.

I often wondered what Sweeney thought of the slaughter of the Indians and the destruction of their way of life. I am sure he was well aware of the parallels with the Irish Catholic experience, but did he have sympathy for them?

He must have led a dangerous and exciting life, however, and one of the tales told about him was that he once was part of a wagon train of merchants who trekked down from Kansas City to Santa Fe, which was under Mexican administration at the time. He survived Indian attacks and buffalo stampedes, and made a small fortune on his trading goods when he arrived in Santa Fe. The trip back was equally dangerous.

When American western movies began to arrive in Donegal in the 1940s, I made many of my young friends very unhappy by telling them that the cowboy and Indian movies were all a big hoax. Based on my father's stories and the books I had read, I told them the real West was not like that at all – that the cowboy era came after the West was won and that the cowboys played only a minor role in the conquest of the territory.

But my friends would have none of my negative attitudes: they were in love with Roy Rogers, Gene Autry, and the Cisco Kid, and they didn't want to hear anything about John Sweeney who had lived up the street and was out in Kansas when the Indians ruled the West.

As I grew older, the urge to go out into the world grew stronger, but I could not decide where I wanted to go and what I wanted to do.

I knew perfectly well that I could not follow in the footsteps of John Sweeney, because both the Indians and the buffalo were gone with the wind; I could not go out to the Pennsylvania coal mines either, even if I wanted to, because they were closed down; and since World War II was over, I could not follow my brothers into battle.

It seemed to me that the world had become a very boring place: there were no wars to be won and no exotic places like the American West waiting to be explored. So I remained tending bar in the family hotel in Dungloe, bored and alienated, and as poor as a church mouse, because business in the hotel was going from bad to worse.

Then, I began to hear rumours about young men from around Dungloe who were making a great deal of money in the Highlands of Scotland, working on hydroelectric schemes. I heard the work was dangerous because it involved driving tunnels through the mountains and constructing dams in the wilderness, but the pay was rumoured to be tremendous – at least four times the normal pay for a labourer – and I found the combination of dangerous work and a great deal of money intriguing.

Of course, the tunnel work did not measure up to fur trapping on the Missouri River or taking part in the Invasion of Normandy, but in an extremely boring world it seemed exciting enough. It certainly seemed better than bartending in Dungloe.

Those who worked on these Highland schemes had acquired the nickname 'Tunnel Tigers', and articles describing their exploits were beginning to appear in the local newspapers.

The more I read about them the more interested I became.

Although I had yet to meet any of these men, I knew that I was hooked at the prospect of becoming one of them and I knew that sooner or later I would head for the Highlands.

CHAPTER I

Tunnel Tigers

I FIRST MET A number of these tunnel tigers in the early Fifties and they made a tremendous impression on me.

I already knew the rumour of big money being earned was not a myth when the children of some of these workers began to show off new clothes, and their wives began to construct new, modern homes at a time when very few even could afford to make improvements on their homes.

The families of the Highland workers had obviously acquired a far better standard of living than I had, because I had not had a new suit in years and I rarely had pocket money.

Business in the bar and hotel had been in decline for a number of years. Our regular customers in the bar were few and far between and none of them were big spenders. The guests for our twenty-room hotel were gradually being lost to the competition, and we did not have the money to bring the hotel up to date.

The decline in our business was in a way a normal process. Most businesses have a life span and they mature at a certain point and then begin to decline. Very few are successful generation after generation.

When my father and mother opened the business after they returned from the United States in the early Twenties, it was a success for several decades, attracting countless customers to the bar and doing exceedingly well with the tourist trade. Indeed, during the summers in the Thirties my parents turned away tourists every night because the hotel was full. The hotel was also a popular venue for wedding receptions and large gatherings to celebrate other events.

But a decline set in in the late Thirties, and then World War II destroyed much of what was left of the tourist trade, and the bedrooms in the hotel remained empty night after night.

By the early Fifties, we were in poor financial shape because there was no steady cash flow coming into the business and the demise of Campbell's Hotel seemed only a matter of time.

Given the general scarcity of cash, you can imagine my surprise on a Saturday night before Christmas when the door of the bar opened and eight prosperous-looking customers walked in. They were to leave more money on the counter in one evening than I had seen in months. All of the men were dressed in new suits, new shirts and ties, new overcoats and shoes, and they looked like they had stepped out of a Hollywood movie.

The group had obviously come from Scotland, judging from the style and colour of their clothes. In the early Fifties, the 'Scotsmen' liked to dress up in bright green, blue or brown suits, which was in sharp contrast to Irishmen, who always dressed as if they were running a funeral home: dark suits and ties; white shirts; black shoes.

The leader of the group was Andy Campbell, who was a distant cousin of mine, and the seven others were an assortment of Campbell brothers, sons and nephews. During the evening, the Campbells ordered double whiskeys, not only for themselves but also for everyone in the bar, and they paid for the whiskey from rolls of money that had to be seen to be believed.

There were four of our regular customers in the bar when the tunnel tigers arrived that evening – customers who had not been drunk in ages because they could not afford the experience – but that night these customers became so intoxicated with free whiskey that they thought they had died and gone to heaven.

The affluence and the generosity of these tunnel tigers impressed me deeply. It was like watching missionaries giving away loaves of bread during a famine.

I was no stranger to migrant workers returning from Scotland with money in their pockets, but the occasional generosity of these workers was nothing compared to the drinks bought when the Campbells came to town.

The migrant workers of previous years had been working on the farms in Scotland picking potatoes or digging drains, and while they made considerable money at this activity, most of it went to support families at home and there was very little left over to splurge in the bars, never mind act like Rockefellers. The Campbells were in the bar almost every day of the next ten, and they continued to treat anyone who happened to be there when they baled up.

The Campbells had some competition, however, when other Highland workers came home for Christmas. These workers wanted to let everyone know about the money that they, too, were earning, and they wanted to emphasise it by throwing money around in every bar in town. But even though the money kept pouring into the till, I was not as happy as I should have been with this windfall.

Apart from the fact that I knew all this new-found prosperity would end after the New Year, I was depressed by the fact that a number of the tunnel tigers coming into the bar were young men I had grown up with and who had been penniless the last time I had seen them – they rarely had the price of a drink, never mind the ability to buy drinks for all the other customers at the bar.

I knew there was a time when these young men would have looked up to me because I was a 'businessman', but now if the truth were told they were far more affluent than I was. They were dressed like millionaires; had no debts; a healthy income; and, unlike, me they had no need to keep up the appearance of prosperity. It was this keeping up appearances that I found the most unacceptable part of my predicament at that time, and it fuelled my determination to leave Dungloe.

Families like ours who had managed to become reasonably successful in business were looked up to by the rest of the community and the assumption was made that hotel owners were rolling in cash.

If the family had also lived in America and came back and bought a business, as my parents had, then the assumption was that these families were very wealthy indeed. None of the families involved would dream of disabusing anyone of this perception – no matter how inaccurate it might be in some instances.

These same families nearly always sent their children off to boarding schools and later on to universities, and this reaffirmed the widespread opinion that they were among the elite in the community.

My own parents had gone into business after returning from America and had sent the oldest of their offspring to university. Because of this they acquired the status of a successful business family. This status was retained even after there was no more money to send me or other younger members of the family away to school.

Being young and critical, I was painfully aware of the gap between the perceived status and the actual reality of our financial situation. And since I often lived in a fantasy world populated by wealthy fur merchants and heroic American troops, the fictional world of the wealthy business person who was in reality very poor seemed to be a very a pale imitation of the exciting life I dreamed about.

Andy Campbell did not help matters by asking me which university I planned to attend. He assumed I would be going, and it obviously did not occur to him that I might have other plans; or no plans at all. However, I avoided the question for the time being and changed the subject as soon as I could. But I was determined to approach Andy about work in the Highlands at a later date.

During the Christmas holidays I came to the decision that there was no future in pretending to be part of a success story, when all that I could see ahead was years of genteel poverty. What was the value of prestige if all my energies were devoted to keeping up appearances?

By the time the Christmas holidays were over I envied those tunnel tigers, who had their freedom and their income, and I made up my mind that I was going to leave home and go where the money was. I was certain this was a good move because I was sure that Andy Campbell would give me a job, since he and my father had been friends for decades.

But I had two reasons for not asking him for a job while he

was at home. First, it was possible that he might refuse me, and that would be embarrassing; second, I did not want to approach him until I talked the matter over with my parents first.

*

My parents were disappointed but not surprised when I let them know about my decision.

My sister Rose was already in Scotland and was working in the Cockburn Hotel in Edinburgh, where she was earning a respectable salary. She also was sending money home every week to our mother to make up for the shortage of money to pay the bills.

My parents knew that every member of the family had left to make a living elsewhere, so there was no reason for them to be convinced that I would not leave like the rest. However, they might have hoped that I would remain at home to take care of them, but they never brought the matter up.

But I felt more than a little guilty because I knew that they would no longer be able to run the business. They just could not manage it on their own.

After they recovered from my announcement, my parents expressed concern about the dangers of working on a construction site. But when I pointed out to them that my brother Bernie had been a radio operator on American tankers during the war and my brother James had joined the American Army in World War II, they agreed that the Highlands seemed less dangerous than war-torn Europe, so I got a reluctant blessing from them.

When they realised that the last of their brood was leaving the business my parents leased the hotel out and reluctantly went into retirement.

The decision to retire was especially tough on my mother, who was sixty but was as much in love with the hotel business as she had been when she was in her twenties. She had great difficulty in accepting that it was all over and that she had to bow out.

I think my father, who was in his eighties, was delighted to be rid of the daily struggle, which he had found less desirable as time went by. Gardening was the love of his life and he had plans to devote more of his time to that.

When they completed arrangements to turn the hotel over to a tenant, they went off to live with my sister Theresa, in Portumna, Galway, and I went off to Scotland.

CHAPTER 2

The Emigrant

GOING TO SCOTLAND to work would not have been my first choice had I any other options open to me at the time. Emigrating to America or Australia would have been preferable but going either route would have required a great deal of money, not only for the fare, but also for living expenses until I found a job. And I didn't have any money at all at this time. In fact I had to borrow the boat fare that would take me to Scotland and the train fare from Glasgow to the Highlands.

Since I had no idea what I would do if I had gone straight to Australia or the United States, Scotland seemed like an ideal place to spend some time, acquiring some assets to finance a greater expedition out into the great big world beyond Donegal.

I had no regrets about leaving Dungloe, although I was sad that my parents seemed upset that I was leaving. But I had a life to lead and like a lot of my ancestors I did not think that Dungloe was a place for young men with ambition. There had to be other places where life could be exciting and productive.

I was excited at the prospect of taking the boat from Derry to Glasgow and stepping out on to foreign soil for the first time. I did not know what lay ahead, but I believed that it would be far more exciting than what I was leaving behind me.

*

The Glasgow ferry left the Derry docks late in the evening and darkness was rapidly closing in as the boat cleared the Foyle Estuary and turned toward Glasgow more than a hundred miles away. As the twinkling lights on the Irish coast vanished beneath the horizon and we plunged into the darkness on a boat that rocked and rolled as it made its way east, I was very much aware that I was making a voyage into the unknown.

When the last light on the Irish coast vanished I knew that I had cut ties with the Irish Catholic incubator that I had been raised in and was heading towards a country that did not subscribe to any of the Irish Nationalist/Irish Catholic belief system that had been very much part of my upbringing. Indeed I had been told by more than one tunnel tiger that the anti-Irish Catholic prejudice in Scotland was exceeded only by the rampant prejudice that existed in Northern Ireland.

But I did not care. I did not care what the Scots thought of me because I wanted to get out of Ireland, and if those young men who had come home at Christmas with their pockets full of money earned in Scotland could put up with this hostility, then I would put up with it too.

Besides, I knew that I was not going to be alone in Scotland – that I would be working with thousands of other Irishmen and there had to be safety in numbers. Anyway, I had just turned nineteen and young men of my age had little fear of anything. Most of the passengers on the boat huddled up in the lounges, stretching out on benches as they tried to get some sleep on a trip that would take all night to complete. I would not see Glasgow until dawn.

I did not want to miss any part of this new adventure, however, so I went out on the deck and made my way up to the bow and watched the progress of the boat, plunging through and over 15-foot waves as it crossed the Irish Sea.

I was no stranger to the sea or to boats. My grandparents lived on Innisfree Island off the Donegal coast and I had spent my summers in the island. Innisfree was three miles off the coast of Dungloe, and I had often made the voyage back and forth on a small yawl that sailed from Dungloe Pier across Dungloe Bay to the island.

I loved those small voyages and thought them very exciting.

But this was a different kind of voyage – although the ferryboat was huge, the open seas tossed it around as if it were a tiny rowboat. Every now and then a giant wave would roll over the bow and spray cold seawater all over me. But I loved the experience.

I stayed on the open deck most of the night, and during that time, I wondered what kind of voyage my father and grandfather had when they went off to Scotland for the first time as young men, in the late nineteenth century. Had the ship rocked like this or had it been smooth sailing? And I wondered how many more of my ancestors had made the same trip for the same reasons as I had.

Eventually, I went below and found a vacant spot on the floor and lay down to try and get some rest. Sleep eluded me on this hard heaving bed, however, and when the grey light of dawn filtered through the portholes, I went back up on the deck again.

*

I had my first glance of the Scottish coast as the Derry boat approached the Clyde Estuary early in the morning.

I absorbed every detail of what I saw. This was the first time that I had ever seen a foreign country and I expected to find major differences from the Irish landscapes I had left behind. But the vistas that presented themselves were of long lines of piers on either side of the estuary and dirty ramshackle commercial buildings lined up at the back of the docks. It was very much like the docks of Derry or Dublin.

I was excited as I came down the gangplank and stepped on Scottish soil for the first time. It is hard to describe the emotion, but it felt like I was embarking on a great adventure. All around me on the pier there were the sounds of Scots accents, mixed with the accents of the Irish who were getting off the boat.

As I disembarked from the boat I tagged along with young men I met on the voyage, as they too were heading for the train station. This saved me the trouble of looking for the station since I did not know my way around.

First we took a train into central Glasgow and then we took another train for the Highlands. I had an hour to spend in Glasgow before I got the train for Perthshire and I used the time to wander around the streets near the train station.

The sights and sounds seemed alien to me. First, there were all the Scottish accents which reinforced the fact that I had left my native soil, and then there were the Africans, Indians and Chinese who strolled through the streets, who were, like me, immigrants from their homelands who had settled down in Glasgow. I had never seen so many Africans and Asians in one location before.

I found all the new sights and sounds exhilarating – I felt that I was on a movie set in a foreign location and that I had a part in it. What a contrast this place was to Main Street, Dungloe, where everyone was white, and almost everyone was Catholic, and a black person was seen only once in a blue moon.

I walked into one of the shops owned by a man who looked Indian and was amazed at the variety of exotic goods that were stacked on the shelves. I was also amazed at the array of smells coming from all the spices and unusual-looking dried meat and fish piled on the counters. Some of the Indian customers were dressed in native clothing, which looked like the Indian version of the Scottish kilt.

Further down the street there was a bar and I decided to walk in and have a drink while I was waiting for the train. But the liquor on display was as alien as the goods in the Indian store, and I ordered a pint of beer of a brand that I had never seen before.

It was early in the morning and yet the bar was full of patrons, and judging by their accents most of them seemed to be members of the Scottish working class.

As I waited for the beer to be poured I looked at the shelves behind the bar and could not see Irish whiskey of any brand on display. Nor was there any Guinness or any of the Irish beers or lagers.

I soon regretted having ordered the beer because it tasted just horrible – sour in fact – and it was not at all like the smooth, velvet taste of the Guinness I was used to. I did my best to drink it all, but I quit after drinking half of it and went back out on the street with a bad taste in my mouth.

Eventually, it was time to leave Glasgow, and as the train made its way through the city heading northwards, I noticed how shabby the housing developments seemed and how poor the decaying buildings in the inner city seemed.

I did not know it at the time, but within Glasgow is an area named The Gorbals, which in the Fifties was a huge Irish slum, a very tough Irish slum, and many of its inhabitants were from my home parish in Donegal, and some of these were my own cousins who had emigrated to Glasgow years before.

Eventually the train eased out of the city and raced through green fields and heavily wooded areas as it made its way northwards.

When we crossed into Perthshire we were officially entering the Highlands, the part of Scotland associated with the clans, the kilts, the bagpipes, and the Scottish Gaelic language.

The Highland scenery was very much like the northwest Donegal scenery that had been my home. Rugged mountains were separated by deep glens and everywhere there were bogs and heather. And here and there were isolated farmhouses hiding among trees.

I noticed also that the place names on the signs were mainly of Gaelic origin and were very similar to the place names I had left behind me in Ireland.

We slipped by small towns as we travelled northwards. They were very similar to Irish towns, except they seemed to be a little tidier – no derelict cars and few rundown houses in sight.

The trip to Pitlochry took two hours, and by noon we pulled into this quaint little town that has been called the gateway to the Highlands. My journey was not over – I still had to take a bus trip to Dalcroy camp, which would be my home away from home while I was in Scotland. I had an hour to wait for the Dalcroy bus, and to pass the time I decided to stroll around Pitlochry. It was a pretty town, full of small hotels and clean, tidy shops and restaurants and neat attractive homes.

On the outskirts of the town a massive dam was beginning to emerge from the ground, and beyond the town an assortment of construction sites were clearly visible. I would have liked to explore a great deal more, but I had no time, so I went back to the bus station to wait for the bus.

The Dalcroy bus, when it arrived, was a dilapidated vehicle that had obviously seen better times. But it managed to stagger off towards the camp anyway.

We set off on a single track road that wound its way up through the glens and mountains towards the camp. As we puttered along every now and then the bus had to pull into a passing place to allow a vehicle to get by, but there was little traffic in either direction.

I eventually saw a large cluster of buildings emerge on the hill-side and a sign on the roadside read – Dalcroy 1 mile. I was arriving at my new abode.

I was not very impressed. The camp at Dalcroy was a complex of large grey huts, all looking alike and all lined up to form streets, each hut with its own number. Two large buildings at the entrance housed a canteen and a mess hall. The camp seemed devoid of personality.

A fleet of rickety buses was lined up near the entrance. I assumed they were used to ferry workers to the construction site. This site, wherever it was, was obviously not in the immediate neighbourhood.

It was early afternoon when I arrived and there were very few people around. I walked about for an hour looking for an office, but could not find one.

Finally, I asked a man who was walking by if he knew where I could find Andy Campbell. The man told me that Andy was up at the office at the construction site at Bohesbic, several miles away. He also told me that buses did not go up there during the day, and if I wanted to see Andy, I had better wait until he came down with the other workers in the evening.

The man went on to say that if I was looking for a job, it was best to speak to Andy privately and not to bother him up on the site. This seemed like a good idea so I remained in the camp.

I decided to explore while I was waiting. My initial impression was that it was like one of those British army camps I had seen in the movies, or one of those German concentration camps I had seen in the newsreels. It was a spartan place.

I was delighted however to see a small building that was being used as a movie theatre. The featured attraction was a Laurel and Hardy movie, with a Roy Rogers movie listed as a coming attraction.

As I wandered around I noticed that the front door of one of the huts was open and I went to take a look inside. The similarity to an army camp was reinforced by the long rows of beds lined up against each wall. The beds were made of black cast iron and the mattresses were no more than three inches thick. Grey sheets and coarse woollen blankets covered each mattress. In the aisle in the middle of the hut were huge cast iron stoves that provided the heating for the hut.

There was one man lying on a bed half way down the hut and when he saw me looking around, he told me to come in and make myself at home. We introduced ourselves. His name was Duffy and he was from Donegal.

'I see by the way you are looking around that you have not been in one of these camps before,' Duffy said.

I told him it was my first time in Scotland.

'You'll get used to it. Everything seems strange now, but in a week nothing will be strange.'

Duffy was friendly and he obviously wanted to be of help. He gave me coupons for dinner that night and breakfast the following morning.

'If you don't have the coupons you won't get anything,' he said.

'Those people down in the canteen would rather throw food away than give it to an Irishman who is hungry. You can have the coupons because I have the flu and I am not hungry.'

Duffy was a shrewd and observant man. Not only did he know this was my first time in a camp he also deduced that I had no background in labouring work, because he said my hands looked soft and my face was not tanned by the outdoors.

'The work is very hard up here,' he said. 'All pick and shovel work. You will be tired at the end of the day.'

Then he asked me if I was a relative of Andy Campbell, since I had the same surname and came from the same part of the county.

Duffy amused me. He belonged to a type that I was all too familiar with in Donegal. He believed that he had to know everything there was to know about each new person he met. Men such as him would keep asking questions until their curiosity was satisfied.

I told him I didn't have a job yet, and I ignored his question about Andy Campbell.

I also decided that I had better get out of there before he weasled out of me that I couldn't go to university and that the family business was failing. I knew that if I hung around there he would know all there was to know about me.

I walked around the camp, checking out every building. There was really very little to see apart from the long line of workers' huts. There was a Post Office, a small clinic, and several buildings that looked like private residences.

I remembered what Duffy had said about everything looking strange to me. He was right. I hoped he was also right when he said I would get used to it.

I had a feeling of being in limbo. The day before, I had abandoned Donegal; today I was in a place that was very difficult to warm up to. It certainly was not the Promised Land.

But even then there was no wish on my part to retreat to the familiar surroundings of Campbell's Hotel. There was no going back and I would endure whatever came down the road.

*

The canteen was open and it had a few customers who had been working the night shift and had come in for a beer before they went to work again.

I had a vast experience in the bar business and I recognised immediately the type of customers who were drinking at that time of day. Judging from their accents they were mainly Scotsmen and judging from the ravished appearance of their faces and their shaky hands they had major problems with alcohol.

I was to learn in the weeks that followed that on weekday

mornings the canteen was mainly patronised by Scotsmen who had a drink problem and who needed a few drinks to cure a hang-over from the previous evening's drinking.

I was also to learn in the years ahead that many Scots had a huge capacity for alcohol consumption – much greater than the Irish had – and they were able to consume this alcohol without letting it interfere with their work.

I got to know several of them who seemed to be at a low level of drunkenness most of the time and yet managed to function very well on the job.

Most of the Irish, on the other hand, could not function at all once they became addicted to alcohol, because they were continuously failing to show up for work and always wound up getting fired. It always amazed me that up in the camps the Scots could drink the Irish under the table without ever getting tagged as a nation of alcoholics.

*

The buses from the construction site arrived late in the evening and I watched as each unloaded droves of workers, each man weary and covered with dirt. They looked like they had been dragged through mud holes all day long.

These were the famed tunnel tigers, and they were not at all like the well-dressed, self-confident rich men that I had seen back in the bar in Dungloe. As I watched them emerge, I wondered what exactly I had got myself into.

I saw a number of familiar faces among those who got off the bus, but Andy Campbell did not make his appearance until the final bus had rolled in. I had no trouble recognising the small compact figure with a hard hat as he stepped off the bus. Unlike most of the others, Campbell's work clothes were clean, and his boots showed little evidence of mud.

I walked over to meet him but was surprised when my friendly greeting was met with a total lack of warmth on his part.

But he listened politely as I asked him for a job – any job – and when I had finished he told me there was nothing at the moment that would be suitable for me. He said in a month or two a job in a store outside the tunnel would become available as the man who had it was not well and was returning to Ireland for good. Until then, the only jobs were those with a pick and shovel at the face.

'I don't think you would want to go in there,' he said.

I knew what he was thinking. He was thinking that I had soft hands and would not be able to cope with the heavy work, and that I had no experience in doing anything but serving drinks in a bar. Of course he was right.

He might also have thought that I had come over to Scotland as an adventure, playing tourist while getting paid for it, and that I was unaware of the nature of the hard work at Dalcroy.

Of course, he had no idea about my financial situation and that the reason I was standing before him was that I needed the money. He had also no idea that I was abandoning the nest for good, and that this job was the first step in my journey out into the world.

Since I had nowhere else to go, I told him I would take a job inside the tunnel, and that I would be very glad to get it. I toyed with the idea of telling him exactly what the situation was back in Ireland, but pride would not let me go that far.

He hesitated for a moment and then pulled a notebook from his pocket and scribbled a message on it. Then he told me to give the note to the timekeeper at the site. After that he turned on his heel and walked away without another word.

I watched him as he strode away toward another end of the camp, where his accommodation must have been. I kept my eyes on the retreating figure until he disappeared into a small building that looked like a private house.

Not once did he look back.

I was astonished at his attitude. Here was someone who had treated me with a great deal of respect back at home only a week ago and had been a close friend of the family all my life. I had expected the same friendly Andy Campbell up there in the Highlands that I knew back at home, but instead he was as cool as if I were a total stranger who was giving him a problem, and I could not understand the reason for it.

However, I was to learn the reason later on that evening when I ran into a man named James O'Donnell, who I had been very friendly with back in Ireland. When I confided in him about Andy's behaviour, he looked at me in amazement and told me I had a lot to learn.

'He can't show friendliness to anyone from home. He must keep his distance from everyone. He is the big boss here, and the bosses can never be friendly with those they give orders to. When you walked in that gate you became just another worker, and that is why he treated you like that.'

He made sense, of course. I was not back at home where being the son of the owners of Campbell's Hotel gave me a certain amount of prestige. By walking away from my hometown, I left that prestige behind me and became just another needy person looking for a job.

My education about the world outside Donegal was beginning that first day in Dalcroy. This was a new world for me and I knew that the only way to survive in it was to put the past behind me and to work as hard as I could, all day every day. Andy Campbell had allowed me to get a toe in the door of this place but the rest would be up to me.

O'Donnell found an empty bed for me in his hut, and then scrounged around for a pair of Wellington boots and a 'donkey jacket' when he learned that I did not have either in my possession.

'You are a real fool. Why did you not bring any working gear with you? You cannot go into the tunnel with ordinary shoes because the tunnel has six inches of water in some places.'

The Wellingtons, knee-high rubber boots favoured by farm labourers, were ideal for the job. The donkey jacket, which was a three-quarter-length overcoat, was lined with felt and it shut out the cold. It was worn by all the workers. A filthy pair of work trousers and a hard hat completed my ensemble.

'You are lucky you met me,' O'Donnell said. 'It is wet and cold in there and if you didn't have boots and a donkey jacket you would not last an hour, even if they let you into the tunnel in the first place.'

As I looked in the mirror that evening I thought I looked a lot like the tattered tramps who came into our bar on a market day. These men always seemed to have a week's stubble on their face and hands and clothes that were black with dirt.

Still, I had the consolation that I would soon be gainfully employed, and no matter how bedraggled my appearance was I would soon be a man with money in his pocket. The donkey jacket might have been old and tattered and the boots had been patched, but I was grateful to O'Donnell for them and I appreciated his guidance.

As I lay there that night listening to the snoring of a score of men, I began to feel a little sorry for myself. But then when I thought of my older brother, James, who had lived in a place like this after joining the American Army and had not only endured primitive living conditions, but had also faced possible death as the army rolled into Germany, I realised that things could be a lot worse.

CHAPTER 3

Bohesbic

I GOT UP EARLY next morning and went down to the canteen, where I got my introduction to camp cuisine. The breakfast menu featured a sloppy mess that was supposed to be oatmeal porridge, accompanied by a platter of ugly yellow lumps that was supposed to be scrambled eggs. Two slices of bread smeared with margarine were also provided. The final item on the menu was a cup of tea that had only the faintest resemblance to the real thing. And yet I ate it all up because I was hungry and because I knew that I could not do a hard day's work on an empty stomach.

The mess hall was all set up with wooden tables and wooden benches. The food was dumped on white enamel plates, badly chipped, and the knives and forks were made of tin and were badly tarnished.

The men and women serving this food at the counters were rumpled looking and seemed bored. They avoided all eye contact with the men they were serving. Needless to say there was no chitchat going on between the workers and the mess staff.

When we were leaving the canteen we were given two sandwiches wrapped in white paper: one was comprised of a sour white cheese spread between two slices of white bread; the other was a sandwich made of Spam, a meat product I had never tasted before. Both looked awful.

'You will get used to it,' said O'Donnell, when he saw me examining the sandwiches on the bus. 'After you eat these sandwiches every day for a week they begin to taste good.'

I doubted very much at the time that I would ever get to like these ugly looking sandwiches, but hunger can do strange things to you and within a week I was eating them up as if they were gourmet food, and I even ate additional sandwiches given to me by other workers.

The bus travelled several miles up the hillside toward the construction site, past flocks of sheep who ignored us, and herds of deer, who kept a wary eye on us.

The construction site at Bohesbic was located on top of an immense mound of boulders and gravel that had been hauled out of the tunnel area and dumped outside. This material had been levelled to form a platform on which all the support buildings for the tunnel operation were located.

The rocks and gravel were brought out of the tunnel by a miniature train that pulled a long string of carriages called skips behind it, and the material was then ferried out to a tip, where it was dumped into a chute at the edge of the tip. The chute fed an immense rock crusher that turned the rocks and gravel into sand. The sand was then mixed with cement to make concrete for the dam located near Pitlochry.

The miniature train engines, called locos, which pulled the skips were also used to ferry workers into the tunnel.

I headed for the timekeeper's office as soon as I arrived and discovered that the timekeeper was someone I knew from Dungloe. I had in fact gone to school with him.

It took me some time to get my papers in order. Afterward I was told to go to the mouth of the tunnel and get a loco into the face and report to the foreman.

As I rode into the tunnel I had three impressions of this new workplace: it was very damp; it was very cold; and it was a very dark hole in the ground.

The loco moved slowly deep into the earth. The project was more than a year old and the face of the tunnel was already more than a mile into the mountain.

A cable of electric bulbs, each bulb fifty yards apart, hung on spikes on the wall, provided the only illumination. Water dripped down constantly from the ceiling and there was a smell of dampness that grew stronger as we made our way toward the face.

As we moved away from the mouth of the tunnel, the entrance, which was at first a large disc of light, became smaller and smaller until it was reduced to a disc the size of a small coin.

Suddenly, I had a strange thought: I wondered if there were rats in this place. I had suffered from a terrible phobia about rats ever since I saw two of them back in the hotel in Donegal late one night after everyone had gone to bed. I had been in the kitchen when I saw the pair of them moving around in the hallway outside the kitchen door.

I had been about to put out the light in the kitchen and go out into the hallway and up the stairs to my bedroom, but the thought of running into these two rodents in the darkened hallway was too much and I stayed in the kitchen until daybreak.

When my father got up in the morning he set two traps and, by the end of the day, the rats had gone to Valhalla.

But the memory of the incident had stayed with me, especially the feeling of being shut in with these rats in a darkened area, and it must have been the dark murky atmosphere of the tunnel that brought back the memory of the rats.

The driver of the loco was a man named Doherty, who was a distant cousin of mine, and I asked him as casually as I could if there were any rats in the tunnel. He paused before answering, and then said the place was crawling with them.

'I hope you have your sandwich in a tin box,' he said. 'Because if you do not, it will be all eaten in an hour.'

I said nothing. I know the Irish sometimes play vicious practical jokes on one another, and I thought he might only be pulling my leg. Besides, what would the rats survive on in a place like this? There was nothing in here but sand and stones. Still, I looked around myself apprehensively. Suddenly, I had more than rats to worry about: there was a tremendous explosion ahead that sent a shock wave and a boiling cloud of smoke and dust towards us. The force of the explosion was unbelievable.

I almost passed out there and then. I thought some terrible accident had occurred and that I was going to be killed in the first hour on the job.

I had no way of knowing then that this was a prank that was sometimes played on newcomers on their first day in the tunnel. The loco driver would time his entrance to be just as the men at the face were about to fire a round of explosives. When the men at the face saw the lights of the loco coming close they would set off the charge. Then all would laugh at the horror that the prank inspired on the newcomer.

Occasionally, a newcomer would jump off the loco in terror and would run all the way outside before anyone could explain to him about the blast, and sometimes, because of fear or embarrassment, he would not return. But I saw the smirk on the face of Doherty, and this ended any inclination I had to make a run for it, because he obviously knew something that I was not aware of. From that smirk I deduced that this explosion was a normal occurrence and not an accident.

Nevertheless, I thought that this was a very vicious prank, and I was annoyed at Doherty, who obviously enjoyed the prank very much.

When I found out later that there were no rats at all in the tunnel, I believed that I owed Doherty some payback, if ever I got the chance to get revenge.

I would get that chance several weeks later, and I can tell you that revenge is sweet.

*

I had been told by the timekeeper to report to a foreman named Griffin, a name I recognised as one of the tunnel tigers who patronised the bar back in Donegal.

When I first met Griffin in our bar back home, he was abusive to other tunnel tigers when the Campbell tunnel bosses were not around, and he treated me like I was his servant. I disliked him very much because Griffin was a very arrogant man and he was immensely proud to be a foreman.

When the timekeeper told me to report to a foreman named

Griffin, I hoped it was some other foreman with the same name. I had no such luck, because there he was as large as life, and when he saw me he stared at me as if I were a ghost.

'What are you doing here,' he said when he recovered from his surprise.

'I am here to work,' I said.

'Well. Well, well,' he said. 'Now we have a bartender on the job. What am I going to do with a bartender?'

I did not respond. I had already learned from my dealings with Andy Campbell the previous day that I could not expect to be treated with any respect on this job by those who were running the project. When I did not respond, he handed me a shovel and told me to start filling the skip with gravel and rocks.

*

There were ten men on the job who operated the drills that bored the holes into the rock face. Other men worked at cleaning up the debris after the explosives had gone off in the holes bored by the drillers.

Fortunately I was not unfamiliar with shovel work, because my father had insisted I help him with the vegetable garden behind the hotel every summer since I was ten.

I had not been involved in this kind of work every day but I had done it often enough that I could handle a shovel without appearing awkward.

Nor was I entirely unfamiliar with hard work – my father had also insisted I help him with harvesting the peat out on a mountain near Dungloe – a chore that was dirty and backbreaking. But I knew that hard work done occasionally was in no way comparable to pick and shovel work all day long, every day.

Griffin watched me as I picked up the shovel and began to move gravel into the skip. I watched the others and saw they shovelled for a while and then worked picking up rocks and throwing them into a skip for another while. They seemed to be pacing themselves.

They did not rush but they did work steadily and there was no slacking off. I did my best to keep pace with them, but it was not easy because my palms soon felt sore from the friction of the shovel handle and my fingers became tender from picking up rocks.

I consoled myself by thinking about the money I was earning on this job – the type of money I could only dream about earning back at home.

We had been working about half an hour when the men who operated the drills began to work. These were key men and they earned the highest rates and the biggest bonuses.

There were only ten of them working on platforms that day, on three different levels in the 29-foot high tunnel.

These men operated long tungsten tipped drills that bored one-inch holes into the rock. When they had bored in twelve feet they pushed foot-long sticks of an explosive called gelignite and a detonator into the boreholes and set off an explosion, which ripped chunks of gravel and rocks out of the face.

Once the explosion had been set off, a machine called an Emco moved in and scooped up bucket-loads of rock and gravel and dumped them into the skips. The work continued non-stop hour after hour.

My job, and that of some of the other workers, was to clean up the loose rocks and dirt that the Emco could not reach. We also extended the rails as the tunnel was carved out of the rock, and did whatever odd jobs needed to be done.

When the face was cleaned up, the mobile platform was pushed forward and the drillers got up on the platform and began the process all over again.

Occasionally a crew got together to try and break the record of the greatest length of footage bored and shipped out in one week. In order to go for such a record every member of the workforce had to work as part of a team, from the crew on the face to the men who worked outside on the tip, because if there were any weak links in the chain, no record would be broken.

Although the bonus system was in place at all times, few crews worked all out week after week. If they did, this would have been very destructive physically for workers like me who worked with pick and shovel. It would just not be possible, so most crews settled for a steady work pace without anyone being put at risk, although there were some 'cowboys' who wanted to go all out all of the time.

The presence of the bonus made it imperative that everyone pulled their weight, and because of this inexperienced workers like me were not that welcome. I knew this and had heard some of these men back in Ireland talk about such 'wasters' – men who did not pull their weight. The last thing in the world I wanted to be was a waster.

That first day in the tunnel seemed the longest day of my life. After about an hour on the job my hands ached, my arms ached and my back ached.

After three hours, just when I believed I could not go on any longer, the foreman called a break. We all sat down on boulders and ate one of the sandwiches we had picked up at the canteen in the morning.

The sandwich tasted delicious even though I knew it was probably the worst sandwich I had ever had. We washed the sandwich down with some hot tea brought in from the outside. Even though the tea had been boiled, it also tasted delicious.

There was very little talk going on between the men. There certainly was no camaraderie. Occasionally there would be an exchange of words, but that was kept to a minimum. Few of the men except O'Donnell and Griffin said a word to me, although I recognised some of them from back home.

Every time I looked over at Griffin I saw him watch me with a smirk on his face. I suppose my exhaustion was written all over me. O'Donnell ignored me for most of the day until I was experiencing another bout of exhaustion and then he came over and said that they would be blasting shortly and we would have a break of more than an hour.

'Just hold on and you will be able to get your second wind,' he said.

I did hold on and I did manage to make it through the ten-hour shift, but at the final whistle I felt that I had just done fifteen rounds with the heavyweight champion of the world – and lost.

That evening I wondered if I was going to be able to hold on to this job because it was the toughest experience I had had in my life.

I knew of course that my problem was that my body was not conditioned to this work, in spite of the fact that I was young and strong. I was barely able to climb onto the bus when we got outside the camp, and I collapsed into the nearest seat. O'Donnell sat beside me and he kept telling me I would get used to the work.

The road back to the camp was unpaved and full of potholes and as I bounced around on the hard metal seats every bone in my body began to ache.

Unlike most of the other workers, the first thing I did when I got off the bus was to go up and take a cold shower.

After that, I managed to make it to the mess for a dinner of soggy potatoes and mushy meat that was supposed to be boiled beef.

I was amazed at how cheerful most of the men seemed to be and how talkative they were. There was none of the silent and driven personalities I saw on the job.

Two of the men I had worked with that day came over and talked to me.

One of them said he knew how I felt – that he almost died from exhaustion the first week on the job two years back.

'But you will get used to it,' he said.

They then invited me to the canteen for a beer, but I was too tired and said no. I instead headed back to my hut.

I was totally preoccupied with my physical condition that night and I could think of nothing else but the ten hours of torture that lay ahead the following day.

In spite of the fact that I was very tired, I had great difficulty getting any restful sleep because the minute I lost consciousness I would have nightmares about rats and explosions. I must have woken up terrified ten times that night. Each time I woke up I would find the rest of the men in the hut snoring soundly. I envied them.

The only consolation I had was that I remembered a story my brother James used to tell about the hardships he went through in basic training after he joined the American Army. He said he thought the training was so tough that he would never be able to hang in there.

But after a week James began to harden physically and emotionally and was able to endure the rigors of army training.

I remembered his story that first night and I made up my mind to tough it out. After all I was not going to have to face armed Germans in the morning – the only enemy I had was my own physical weakness, and I was determined not to show any lack of strength no matter how weak I felt.

In a way I was very fortunate that I had a considerable amount of pride because it was this pride that enabled me to endure whatever came down the road. Without it I would have been on the next bus out of Dalcroy.

CHAPTER 4

The First Wages

ON MY FIRST DAY in the Highlands I was told that I would adapt to the food and the living conditions, and I did – within a week. By then I no longer turned my nose up at the unappetising meals and I felt at home in the crowded hut.

Indeed, after the first week I looked at the camp as my new place of residence and the hut as my new home. The hut was warm, the bed was comfortable, and I was surrounded by friendly people.

I was surprised how quickly I adapted to the work inside the tunnel, because on the evening of the first day I thought I might be dead within a week. Although I was still bone weary at the end of the day during the first week, by the end of the second week I had no trouble at all holding my own. By then, my muscles were getting toned and the palms of my hands were getting used to the friction of the shovel.

I think Griffin and the others were just as surprised as I was by how quickly I adapted. O'Donnell told me later that they were convinced I would never stick it out, and they were talking behind my back, making bets about how long I would last.

I knew I was being closely monitored the first few days I was on the job, because every time I looked up from my chores I could see that someone was watching me. This only increased my determination to stick it out.

I knew the other men had no confidence in the ability of 'college men' – anyone who graduated from high school – to do a hard day's work. Back in the bar I had been told often enough by a drunken labourer that a hard day's work would kill me and that all I was good at was reading books and pouring drinks. They would apologise next time they were in for being insulting, but I knew that was their opinion – they thought white collar people were wasters.

Some of my co-workers had difficulty understanding why I was there in the first place. They didn't know that I had no option but to stick it out, and they did not know that pride would not have let me walk away no matter how tough the job was.

The truth was that I would have been humiliated if I walked away from the job, and I knew that word of my desertion would get back to Dungloe in a hurry.

*

I found the explosions that went off in the tunnel to be the most difficult thing to get used to. Even with my ears covered the whiplash crack of the disintegrating rock and the gush of smoke and dust was overwhelming.

And although we were well back from the face when the explosion occurred, occasionally a rock flew in our direction, or a chunk of rock was loosened from the ceiling with the concussion, and this made the event all the more dramatic.

The highlight of my first two weeks was my first wage packet. It was four times what I would have made as a clerk back home. It was paid in cash in an envelope on a Friday afternoon.

I counted the cash carefully when I received it and put it away in an inner pocket. I felt like taking the envelope out again and again to count the money but I resisted the temptation, because none of the others did that and I did not want to look like a novice.

The money provided me with a warm glow and feeling of accomplishment, since this was the first time that I could say the money was mine, earned by me, and nobody else owned a share of it. For the first time I had assets.

Later on that night I lay in bed writing a letter to my parents. I noticed that many of the other men, who were also writing letters, were slipping money into the envelopes. After thinking about my elderly parents back at home living from hand to mouth I put ten pounds in with my letter, too.

It was a practice I was to continue with while I was in Scotland,

and I felt it was the right thing to do. I had been close to my parents and the thought of not sharing would have made me feel guilty.

As I got to know the men better in the months ahead I became aware that almost everybody on the site was sending money back to Ireland to someone – wives, mothers, or children. Most of them were in fact there not just to support themselves but to support others as well.

When I had settled into the job, I decided that if I was going to be marooned up in the Highlands for a year or more I might as well make as much money as I could, so I volunteered for overtime work on Sundays.

The type of work that was available on weekends was maintaining the tracks inside the tunnel; poking the roofs and sides of the tunnel for loose rocks to bring them down before they fell down; and changing light bulbs in the miles-long electric cable.

Although it was easy work and only a few men were needed it was sometimes hard to get men to work on weekends. Many of them wanted to head for relaxation in Perth and Glasgow.

Since I was always available, I very often got to work on weekends, although Griffin was reluctant to give me this work because he still held on to the idea that I was an amateur, even after a month on the job. But I persisted and got the jobs anyway.

I wanted the weekend job for another reason. Jimmy Doherty, who was driving the little loco that ferried men and material in and out of the tunnel, owed me a favour because he had set me up for the explosion that first day on the job. I intended to collect the favour. On the weekends that I worked, I harassed Doherty into teaching me how to drive the loco.

At first he had dozens of excuses why he should not do this – it was against the law; I might kill myself; I might kill him; he would get fired if he were caught showing me; I would get fired for driving the loco; Andy Campbell wouldn't like it.

He went on and on, but I persisted and wore him down. I knew the real reason he did not want me to drive the loco was that he liked to tell everyone how difficult the job was and he did not want me to find out how easy it was.

So I continued to bother him until he gave in and taught me. Then, it became a routine every Sunday for me to take over the little loco, and I just loved to drive it in and out of the tunnel.

*

One evening when I was on the job about eight weeks, I saw Andy Campbell standing among a group of men waiting for the bus to take us back to camp. I had not seen him since the first day at the camp and had made no effort to contact him after O'Donnell had set me straight on the relationship between a superintendent and a tunnel labourer.

Campbell stood to one side and let other men board the bus, and then when I came up to the bus door, he walked in behind me and sat down next to me. I was very surprised but did not say anything.

After the bus got under way he told me that the job in the store was available and that I could start the following Monday.

I picked my words very carefully. I told him I appreciated the offer, but the store only paid half of what the job at the face paid and that if he didn't mind I would stay where I was. I thanked him for giving me the job I already had and said I would always remember his help.

Then I told him I would be interested in driving the loco if such a job ever came up. I told him I enjoyed driving the loco and that I would do a good job if given the opportunity.

He said he would see what he could do.

'I hear Doherty has been teaching you,' he said. 'I will think about it. There isn't one of those jobs open now, but I will see if something might come up at one of the other sites.'

Campbell was silent for a while, and then he told me that he had also heard that I had been in the canteen every Friday and Saturday night, and sometimes on Wednesday.

He said it was not good for someone so young to be around bars like that, as whisky could get a hold on my mind.

I felt like laughing. Here was a man who drank a bottle of whiskey at one sitting over in Donegal and now he was lecturing me.

But in my heart I knew he was only talking to me for my own good and was showing his concern. Besides, he had been drinking when he was on his holidays, and I had heard that he was a very moderate drinker while in Scotland.

I did not take offence at him and I told him that I drank beer not whisky and that I never missed a day's work.

'Beer can get a hold of you, too,' he said. 'Don't get too fond of it. You will become a martyr to it.'

I said I would keep an eye on myself, and he need not worry. But the way I modified my behaviour was to stay out of the canteen on weekdays, and go off to the bars in Dunkeld or Pitlochry on most weekends.

I suppose I really did not fool Campbell with this strategy, because he obviously had informers who were keeping him briefed on who did what. I resented these informers, whoever they were, and thought them a bunch of old ladies who liked to spread gossip.

Two weeks after my chat with Andy, Doherty had a big fight with Griffin over an issue that had nothing to do with me, and Doherty walked off the job.

Doherty had not reached the mouth of the tunnel before I had asked Griffin for the job, telling him I could keep the loco running and he would not even miss an hour's production.

Griffin, too, had heard about Doherty teaching me how to drive the loco, although he never mentioned it to me.

Griffin hated activities like that going on behind his back and he would have read the riot act to both of us were it not for the fact that Doherty and I were both cousins of Andy Campbell.

Griffin hesitated when I asked him for the job. He really did not like me from the first day and he would never have given me the loco job if he had had any other options. I do not know why he did not like me, but maybe the reason was that he knew I did not like him either.

Anyway, the only other person who could drive the loco was a middle-aged man who wanted nothing to do with the job.

The man had been drunk on the job on another construction site and almost lost his life when he ran a string of skips off the end of the rail line. He had lost his nerve after that and never drove a loco again.

So, Griffin reluctantly agreed to try me out, but he said he was going to ride with me on the loco for a couple of runs before he allowed me to have the job.

I was very nervous as I made the first run because he was paying critical attention to every move. But the machine was not complicated – there was a lever which controlled the speed and a pedal that controlled the brake. However, as the loco picked up speed on the surface, which was essentially level, the heavily loaded skips took on a momentum of their own, pushing the loco ahead, and when this happened you had to cut the power from the loco until the momentum slowed. Then you fed the motor power again.

The danger of ignoring the momentum created by the skips was that the driver was no longer in control, and if you came to a bend without being in control, the loco could jump the tracks, as the brake would be useless as a means of slowing the loco because of the combined weight of the loco and the loaded skips.

Jumping the track at a bend was the most common type of accident involving a loco and the accident usually involved an inexperienced driver.

I knew every yard of the track by this time and I knew when to give power to the motor, and when I could coast along.

Along the way I slowed down before every bend and crawled around them. When I reached the end of the line, I just inched carefully forward.

Griffin was impressed. He said I did well. He said I really surprised him and that I was not nearly as stupid as he thought I was. He said I could have the job. I said – thank you very much.

I really liked the job on the loco, although I was still working for Griffin, and he was just as abrasive when I was on the loco as he had been when I worked with a shovel under his eye. But I had learned to ignore Griffin's insults, and I hoped that one day I would get a chance to pay him back.

I knew, however, that I could never confront Griffin, because if I did I would be the loser. Griffin was a very valuable asset to the company since he kept production moving and did not allow any wasters on the job. If you did not do your work, you were out of the tunnel no matter who you were. And that was good for the company.

*

I do not know what the others thought of my rapid 'promotion'. Nobody said anything to me about it, but I am sure that they thought I was in a great hurry to get away from the hard work at the tunnel face and get a soft job on the loco instead.

Part of this was true, of course. But I had worked hard as a labourer and gave no-one room to criticise me, and I had been smart enough and ambitious enough to learn to drive the loco, and this eventually led to the job

So, I did not feel guilty about getting the new job or the increase in pay that went with it.

Anyway, I had absolutely no qualms about trying to better my situation. It was very tough in the tunnel, and I believed that anyone who had the chance to make life a little easier should reach out for it.

A week after I got the job, Doherty came back on the site and wanted his job back. Griffin refused him. Then he went to Andy Campbell, and Campbell also said no, but he gave him a pick and shovel job on another site.

I met Doherty in the canteen one night and he accused me of taking his job. I said he quit his job and he should blame no one but himself.

We rarely spoke to one another after that.

CHAPTER 5

A Dangerous Place

I KNEW THAT TUNNEL work was dangerous. There were threats to life and limb everywhere. Falling rocks and live electric cables were a constant threat, and the heavy machinery operating in a confined space posed an additional hazard. But I viewed the threat of an accidental explosion inside the tunnel as the most serious one, although I never mentioned it to anyone, and none of the other workers expressed any concern about this threat to me.

There had been no fatalities in the tunnel while I was there, but I knew that four men had been killed near Pitlochry when some explosives detonated prematurely, blowing them to pieces. Several men had also been killed in the construction of a six hundred-foot shaft a mile away from the tunnel.

The shaft was being bored from the bottom up. Each day the men had to climb up on ladders to drill the holes for the explosives. When the holes had been bored they hid in niches cut into the side of the shaft and set off the explosions.

Each explosion sent tons of rock showering down the shaft. Sometimes, a rock flew into the niche, killing a worker, or the explosions damaged the ladders, and when the men went out on the ladders, the ladders gave way and the men crashed to their death.

A fear of explosions had been with me for a long time. When I was eleven years old a score of young men from my area discovered a German mine floating in the water near a beach about five miles from Dungloe.

Being full of curiosity and mischief, they threw a rope around it and tried to haul it up the beach. All of the men involved knew it was a mine and that it was extremely dangerous, but they continued to haul it ashore anyway.

They had just got it out of the water when it exploded. Many of them were blown to pieces; others were badly maimed.

I was a child and was playing out on the street in Dungloe when the explosion occurred. I heard this tremendous boom, which shook windows. People came out on the street to find out what had happened.

Prior to that day we had occasionally heard booms coming from far out at sea as British warships dropped depth charges close to the Irish coast, trying to destroy German submarines that were lurking there.

Sometimes a naval battle between the Germans and the British would erupt out at sea, which could be heard along the Donegal Coast. But we did not witness any battles close up, and the nearest any of us got to the hostilities was when we heard these distant booms, or when the debris from sunken ships was washed ashore.

This boom was different, however. It was loud and seemed very close, and we were sure that something had happened very near to the town.

Shortly afterwards, when word reached Dungloe about the cause of the disaster, my father went down to the area to see if he could help. The scenes that he described later that night of dismembered bodies gave me nightmares for years.

The word 'explosion' took on a new meaning for me after that. It was not just a very loud sound that made your ears pop. It was a murderous force that could decapitate and blow a human body to pieces. So, I was a little apprehensive about the explosions taking place in the tunnel because I had been made aware of what explosives could do.

However, I knew there was a major difference between a mine, which was designed to detonate at the least provocation, and the sticks of gelignite used in the tunnel, which could only explode if a detonator was first set off inside a chunk of the explosive.

Indeed, gelignite could be burned, dropped, or run over with a truck and nothing would happen. It was completely safe, unless a detonator set it off.

Still, an explosion could go off prematurely, like the accident at Pitlochry, which was caused by a lightening strike on the outside electrical cables which were linked to the detonator. The explosion at Pitlochry was a very unusual occurrence, however, and was an example of very bad luck. The men in the tunnel knew the chances of an accident were remote, providing everyone was safety conscious. But no one could prevent an 'Act of God'.

It was a consolation to me that accidents inside the tunnel were rare. Still, I knew accidents did happen inside from time to time, sometimes with fatal consequences. The most common accident in the tunnels occurred when, unknown to the workers, a detonator failed to explode inside a borehole.

When a drill was then inserted in the hole and the new drilling began, the detonator went off, which exploded the unused gelignite, with tragic consequences for anyone nearby. Accidents like that sometimes happened when the men were trying to break footage records and were careless about checking all bore holes carefully because they were in a hurry.

*

I did not witness any fatal accidents in the Highlands, although I did see some close calls. One Saturday when I was not working I took a motorcycle trip with a German friend to another hydroelectric site at Garve, Ross-shire, to visit some friends who were working there.

We were standing at the mouth of the tunnel when a long line of skips rolled out and came to a stop in front of us. The skips were destined for a tip at the bottom of a steep incline.

A wedge was placed under the front wheels of the first skip to prevent the skips from moving, since all the skips were on an incline. Then the loco was unhooked and was shunted off to a side track and a cable was attached to the last skip.

The idea was to ease the train of skips down a very long, steep incline to a dump at the bottom, and the cable was fed out of a large drum, which was controlled by a motor.

This chore had been done several times a day for years without incident. But this time, while we watched, the cable snapped when the skips were only fifty yards down the hill, and the line of skips went thundering down the incline out of control.

There were three workmen and the driver of a bulldozer at the bottom of the hill. Fortunately, they saw the skips coming and ran away from the danger zone. But as they ran the skips derailed and an avalanche of skips, boulders and gravel careened down, barely missing the men who were fleeing for their lives.

It was a close call, and one of the men quit on the spot because he was unwilling to work in that location again.

We had also been put in harm's way by the accident, because when the cable broke, it lashed backwards over our heads and decapitated a chimney on the hut near where we were standing.

But it happened so fast that we did not even have time to feel afraid.

*

There were some close calls at Bohesbic while I was there. On one occasion the rail lines had been laid too close to the edge of the tip, and a loco and a line of skips that were travelling just a little too fast slid over the side and went rolling down to the stream two hundred feet below. No one was hurt.

After I took over the job of loco driver, I was driving out of the tunnel one day when a chunk of rock, six feet long and several feet wide dropped from the ceiling fifty feet ahead of me. I was barely able to stop the loco from running in to it, and I was grateful that I had not being going a little faster or the rock might have dropped on me.

On yet another occasion, I almost ran into an electrician named Gordon who was lying unconscious on the tracks. I did not see him until I was almost on top of him. I had been coming out of the tunnel to get a bucket of tea for the few men who were working there, and I was not expecting to see anyone, because I did not think any of the electricians were working that Saturday. Fortunately for

Gordon, and for me, I was keeping my eyes on the track ahead and saw him in time.

Gordon had been working on a cable when he got a jolt that knocked him unconscious and threw him onto the tracks, and he lay there while I made my way out towards him. But I was able to revive him and ferry him outside. He survived.

All these incidents did not make me afraid to work at Bohesbic but they made me aware of the need for great care. The unexpected could happen anytime.

However, it was the threat of an explosion in the dark spooky tunnel that was always at the top of my mind, and I could never really get used to the savage sound of the detonation, like a giant whiplash, and the accompanying rush of smoke through the dark cavern.

There was no accidental explosion in the tunnel while I was there, so my fears were unfounded.

However, forty years later – almost a lifetime later – in another country – the United States – in 1993, I was working on the sixty-third floor of One World Trade Center in New York, when the building shook and I heard the familiar whiplash crack of an explosion – that instantly brought me back to Bohesbic.

Later, as I went down into the basement of the World Trade Center and viewed the hundred-foot-wide, seventy-foot-deep hole a terrorist bomb had made, the smoke and the darkness of the basement transported me back forty years to the tunnels in the Highlands of Scotland.

Nowadays, when I hear people say that they have forgotten certain incidents in their life I am inclined to think that important incidents are not really forgotten – that they are tucked away deep in the memory and they can be recalled in living colour under certain circumstances.

CHAPTER 6

The Grand Scheme

WHEN I FIRST ARRIVED at Pitlochry, I knew very little about the extent of the hydroelectric scheme in the area; I thought that the entire project consisted of a dam at Pitlochry and a tunnel up at Bohesbic. But as time went by I became aware that there were other dams, there were other reservoirs, and there were other tunnels and shafts being constructed all over the area.

Indeed, this hydroelectric scheme was the most ambitious scheme of its type ever undertaken in Britain. In all, two reservoirs, nine dams and nine power stations were created, and numerous tunnels and shafts were driven through the mountains to shepherd the water from a dozen lochs and rivers into the turbines, which in turn generated the electricity needed to give power to British cities.

I was utterly amazed at the immensity of the project and astounded that everything was on a grand scale. The tunnels were almost thirty feet in diameter, and the dams and reservoirs were easily the largest man-made projects I had ever seen. Even the trucks used to haul the gravel were huge and there were fleets of them around.

Then there were the huge excavators used to move the rocks and gravel around. I had never seen excavators before and I was amazed at their power and all the work they could do. And then there were the massive bulldozers that pushed gravel and huge rocks ahead of them. I never got tired watching these bulldozers work.

I was proud to be working on such a massive project and I knew that no project of this scale was underway in Ireland at this time. Indeed, the hydroelectric scheme at Pitlochry became an example of British power to me, proving that Britain was still a powerful nation even in the immediate aftermath of World War II. The resources needed to create this project were obviously immense, and the fact that the British had these resources even after fighting World War II was amazing.

I was impressed also that the British Government was involved in such long-range planning and construction. The British were using these resources as an investment in the future, not as a vast public works project to create new jobs for people like me.

The payoff for the British would be the availability of cheap electricity and this benefit was clearly evident to the businessmen in Britain's major cities who were in favour of these projects.

The British first became aware of the importance of ensuring a reliable supply of nationally-generated hydroelectric power after World War I made it difficult to get the needed supply of oil into the country to operate oil-driven electric generators. Britain had no oil of her own at that time.

The need became even more important during World War II, when Germany cut off much of Britain's oil supply in the Persian Gulf, and American oil tankers had to run a gauntlet of Nazi submarines when crossing the Atlantic to get oil to Britain.

After World War I, the British began to build hydroelectric plants in Scotland, and after World War II the pace of construction accelerated. All of Great Britain was surveyed, and the most ideal locations were usually found in the Highlands.

The components of most of these projects included a long deep glen to hold a massive amount of water; a concrete dam to control the water; and tunnels driven through the mountains to bring additional water from adjacent lochs into the glen. The water was then run through turbines set in a dam and electricity was generated.

These massive projects took years to plan and they required hundreds of pieces of equipment, which were brought into remote areas for the machine operators who would operate them. Only a nation like Britain, with its considerable wealth, could finance projects like these.

In addition to the equipment needed for these projects, an army of men was also needed to get the projects started.

Camps for the men were erected in remote areas near the construction sites, and all the services that were part of the camps, such as living accommodations, stores, mess halls and canteens, had to be erected before the first workers arrived.

The recruitment of labourers was a major problem that faced the construction companies, because it took a very special worker to endure the working and living conditions on these projects.

Recruiting labourers from among the Scottish working class proved difficult. The Scots did not have a tradition of working long hours on construction sites that were located a long way from home. Most of their experience was on construction sites within commuting distance, or in coal mines located in their own neighbourhoods.

Nor had the Scots acquired a taste for the dangerous and backbreaking work that was involved in building dams and driving tunnels through mountains.

And they certainly had no love for living in huts with forty beds, sharing the night and the body odour of total strangers, many of whom were Irish. It was these living conditions that were the greatest deterrent to getting Scots labourers to work on the schemes in the Highlands

So recruiting the Scots labourers for these projects was a very hard sell and was seldom successful. There were, however, skilled Scots tradesmen on the job; welders, plumbers and carpenters, known as the Black Gang, who were enticed up to the Highlands by the high wages and much better living conditions than the labourers enjoyed. The better living conditions included cubicles that were designed for two occupants and were reasonably spacious, and shower rooms that had to be shared by fewer workers. The members of the Black Gang were also served better food, and they received a number of other perks designed to keep them on the payroll. Even so, it was still difficult to get Scotsmen who were top rated tradesmen.

The few Scotsmen who had signed up as labourers were usually from a string of islands off the West of Scotland, where Gaelic was spoken. Many of these islanders spoke Gaelic, and some were Catholic, and their culture was in many ways similar to the culture of the Donegal Irish.

These men shared with the Irish an ability to endure the most brutal working conditions as long as there was a hefty pay packet at the end of the week.

When the Highland projects were in full swing only a few Welsh or English workers could be found on the workforce at Dalcroy, and even those few rarely hung around very long, and soon drifted off to more civilised locations.

The reputation of the camps as wild, unruly places full of drunks, gamblers and prostitutes also hurt recruitment efforts.

There was a widespread conviction that any law-abiding Briton took his life in his hands by entering such a place. If he did not get murdered or maimed the first day he set foot in a camp, it would happen to him shortly afterwards. The fact that body bags were not coming out of the camps on a daily basis did nothing to dispel the myth.

Part of the reason for the myths surrounding the camps was the fact that when the Irish workers and the workers from the Hebrides went to Perth or Glasgow on weekends, they were inclined to behave like sailors on shore leave in a foreign port after an extended voyage, or cowboys entering Dodge City after a long cattle drive. The men were thirsty for liquor and excitement and they made the staid Lowland Scots a little uptight. It was widely believed that the workers behaved like that every day of the week, which was not true.

The fact that the majority of the tunnel workers were Irish Catholics only added to the negative image the camps produced among the Scots.

But the reality of the camps during the week was that the workers led a life that was a daily grind of poor food and hard work and the vast majority of them went straight to bed every evening after work, too tired to drink, fight, or fool around. There was no violence and no heavy drinking.

Only on the weekends did the workers enjoy a little relaxation and even then they did their carousing outside the camp. Indeed, the average Scots worker, who was so afraid of these camps, was far safer inside the camps than he was in the streets of Perth or Glasgow, where muggings and other violence were commonplace.

However, in spite of the unwillingness of the average British worker to be recruited for these projects, construction companies like Carmichael had no problem finding workers to bore tunnels in the mountains or erect huge dams in the wilderness.

All the executives of the Carmichael Company had to do was to let the Irish know that another major project was in the works, and the Irish would descend on the area in droves, bringing with them strong backs and a thirst for adventure.

In no time at all Carmichael would have his tunnel tigers in basic training.

*

The Irish were attracted to the Carmichael camps for a number of reasons. First of all Carmichael paid all classes of workers top wages, and, in addition, the company paid a bonus based on production.

On top of that, Carmichael put Irishmen in key spots in management, which made the Irish feel that they were getting fair treatment. Once these benefits became known, Carmichael had no problem in attracting the right kind of Irish labour.

But the money and other benefits were not the only reason that the Irish flocked to dangerous work like this.

There was a tradition among certain segments of the Irish working class of being itinerant labour mercenaries who were always willing to take on the most dangerous chores in far off places, if the pay was well above average and the working conditions were considered harsh by most other workers.

In a way these Irish workers considered themselves a 'foreign legion' of adventurers who had a thirst for new challenges. They liked the excitement, the challenge, and the money, and it gave them bragging rights when they were in the company of workers of other nationalities who had no stomach for such work.

In the nineteenth century, the Irish and Chinese dug the great canals across the north eastern United States, and after that they built the railroads from California to New York. When the coal mines of Pennsylvania and Ohio were opened up, the Irish were among the first to go into the deep pits.

And when gold was discovered out West, they grabbed their picks and shovels and headed for the frontier. Many of them made fortunes in such places as Virginia City, Nevada, while others just made big bonuses, which they threw away on gambling.

In the twentieth century the Irish worked as 'sand hogs' beneath the streets of London and New York, carving out tunnels for water and for subways.

And they played prominent roles in the construction of the Holland and Lincoln Tunnels that link New York and New Jersey.

So it was little wonder, when the call went out from the Carmichael Construction Company in the 1940s and 1950s for men to work on a dam and a tunnel in the mountains above Pitlochry, Perthshire, that Irish sand hogs and tunnel tigers came in from all over the world to get involved in the new project.

In a way, the British were very lucky to have the Irish on hand to drive their tunnels and build their dams. If the Irish had not been around, who would have worked on these projects?

CHAPTER 7

Opposition to the Grand Scheme

IN SPITE OF THE widely acknowledged benefits of the hydroelectric schemes, there were many people living in Pitlochry and in small towns adjacent to the construction sites who were opposed to the whole idea.

These people believed the project would destroy the beauty of the Highland landscapes, and this in turn would destroy the thriving tourist industry. If the tourist industry was destroyed, then the livelihood of thousands would be affected, and the people of the Pitlochry area thought that the benefits of an abundance of cheap electricity were not a suitable compensation for this loss.

So, when the Tummel Garry Hydro-Electric Scheme was proposed, mass meetings were called and protests were directed at the North of Scotland Hydro-Electric Board, which managed the project. But all to no avail. Since the national benefits seemed to far outweigh the local drawbacks, the scheme went ahead as planned.

The Countess of Airlee laid the foundation stone for the Pitlochry Dam on 25 April 1946. In the years that followed each of the other elements in this massive project was commenced in a carefully co-ordinated schedule, as new roads, tunnels and reservoirs appeared on the landscape, one by one.

When Carmichael won the contract for tunnels and a surge shaft in the Pitlochry area, he hired Andy Campbell, who already had a reputation as a 'tunnel tiger', and told him to hire 400 good men who were not afraid of tough, dangerous work.

Andy knew where to find these good men, because he had worked with them all over the North of Scotland on other projects during the previous twenty years.

Andy put out the word in north west Donegal, and soon tunnel tigers were pouring in from all corners of the world.

As the work progressed some of the worst fears of the tourist industry and the Pitlochry environment lobby seemed to be realised. Giant machines carved up the countryside and dug deep trenches that were to be the foundations for dams and reservoirs.

The picturesque mountains, home of deer, sheep, fox and other wildlife, were carved open and vast amounts of boulders and gravel spewed out of the wounds as tunnels were driven deep into their sides.

The drilling machines and the fleets of trucks made a terrible racket all day and well into the evening. Periodically, tremendous explosions ripped through the virgin Highland wilderness, which had not heard such a commotion since the Battle of Killiecrankie in 1689.

And as if this were not enough to disturb the life of the easy going Scots people who lived in the area, hordes of foreign workers, including the Irish, descended on the district. By their sheer numbers they made some local people a minority in their own area.

It must have seemed to many people living in Pitlochry and other nearby towns that they had lost control of their own mountains and glens. Up until this point, Pitlochry had been like an upscale resort or national park where the middle and upper classes came to enjoy the natural beauty of the area and these visitors included members of the British aristocracy, including the King, who came to spend extended holidays in the Highlands.

The Highlanders were used to such people and knew what to expect from them, and they had gone out of their way to make them welcome, since tourism was such a very important part of the economy of the area. As a result the area had developed better class stores, restaurants and hotels to cater to those visitors.

But the multitude of labourers with their alien accents and alien ways were another matter. No welcome mat was out for them. Besides, these workers never patronised the fancy hotels or restaurants, and certainly did not shop in the better stores.

When I arrived in Pitlochry, the local landscape had already been considerably chewed up and the beauty of the surrounding hills showed signs of molestation by a horde of construction workers. There was a gigantic 80-foot pit near the town, which was the site of the Pitlochry Dam, and the topsoil all around the town had been disturbed.

Yet there were still numerous tourists in Pitlochry even though the town was on the edge of a vast construction site, and royalty still visited the area.

While I was there Crown Prince Akhito of Japan visited the town and was greeted by hundreds of protestors who were angry at the atrocities committed by Japan during the war. Strangely enough there were scores of German ex-POWs working up in the hills, some of them still ardent Nazis, but no one bothered to protest about them.

The disruption of the landscapes around Pitlochry went on for years, but eventually all the projects were completed and gradually grass, shrubs and trees grew over the areas that had been disturbed and the natural beauty of the area was restored.

The most remarkable aspect of the completed projects was that the immensity of some of them was hidden completely from the human eye.

The vast 29-foot tunnels were buried underground and could not be seen, and the sites that were a beehive of activity during construction were buried under floodwaters when the project was completed.

The Pitlochry Dam, which seemed a gargantuan project when it was being erected, now seemed small in scale.

The reason for this is that 80% of the project is underground, with the foundations sixty feet below the water level. Thus, the dam is like an iceberg that can show a deceptively small tip, but which has a massive bulk hidden from view beneath the water

Very few of the visitors to Pitlochry today have any idea of the immensity of the scheme that dominated life in the area for more than ten years.

Only the locals know that a vast loch that is now a playground for all to use was created by the dams and reservoirs. And only the locals know that there are areas that disappeared beneath the waters as they covered the landscape. Indeed the recovery is almost complete and the only evidence of the project is the ugly power pylons that march across the glens and hillsides, creating an eyesore that mars the natural beauty of the place.

There is also very little evidence that an army of foreign workers marched through here and lived in those hills for years. The dams, tunnels and reservoirs are the only monuments to the fruits of their labours.

The Irish have an additional monument that reminds local people that they were once there in great numbers. This monument is in the form of a church, St Bride's, in Pitlochry; a Roman Catholic chapel that was built by the Irish workers for their own use. Built with their labour and financial support it owes its very existence to the Irish who worked in this area.

St Bride's belongs to a list of hundreds of Catholic Churches across Britain, Canada, Australia, New Zealand and the United States that were created by Irish labourers with their pennies and their labour. These churches were used by the Irish for only a brief time until they moved on to greener pastures, leaving the churches for the use of other immigrants.

One of the oldest Catholic cathedrals in the United States, Old St Patrick's on Mott Street in New York, was built at the turn of the Nineteenth Century by Irish immigrants, but now serves Chinese and Hispanic immigrants.

St Bride's stands in the Highlands as a monument to the departed Irish and now serves tourists from all over the world instead of the Irish tunnel tigers from Donegal.

CHAPTER 8

Out of the Tunnel

DURING THE FIRST WEEKS on the loco job I enjoyed the work for a number of reasons. First of all it involved some skill, which was more than could be said for pick and shovel work. Above all it did not involve that back-breaking work with the shovel, which put calluses on my hands. The new job also paid more money, and it involved far less physical stress than the job at the face.

I liked the long trips in and out of the tunnel because I was in control of my own little vehicle, and at first I thought I had found a job that I could be happy with as long as I worked up in the Highlands.

However, as the weeks rolled by, a health problem emerged that made it very difficult for me to continue working in the tunnel and forced me to try and find another job outside.

It was in early spring and the weather had undergone a change – the cool Scottish weather had become very cold, plunging down below freezing, and I picked up a cold from shuttling back and forth between the relatively hot tunnel and the freezing atmosphere outside. When it was not freezing outside it was pouring rain.

In spite of the cold I worked every day, even though the first cold was followed by an attack of bronchitis. Bronchitis was something I had been plagued with since I was twelve, so it did not take me by surprise.

I shook off the cold after two weeks, but a week after that I got another cold, and again it was followed by another bout of bronchitis, which sapped my energy and gave me the hacking cough of a chronic invalid.

Three weeks later I went through the same routine. At this point I decided that I had better either go back to labouring at the face or find a job outside, as this running in and out of the tunnel was killing me.

As luck would have it I heard that there was a job opening up outside on the tip, and I decided to go for it.

The job involved being a stock keeper for the explosives, detonators, and the cable used for blasting at the face. All the explosives were stored in a small shack located well away from the rest of the tunnel facilities, and I learned from the Scotsman who was leaving the job that there was no work involved and he was leaving it because he was bored.

By this time I was stressed out from hard work on the face and poor health on the loco, and the whole idea of a job that involved little work seemed perfect for me. I doubted very much if I would be bored.

So, I approached the head stock keeper on the site and I told him I was interested in the 'powder monkey' position, which was how the job was described.

Grimes, the stock keeper, was an alcoholic Scotsman whose face and disposition were ruined by excessive drinking. He hated everybody.

When I asked him for the job he looked at me as if he had not heard me correctly. Then he dismissed me with a wave of his hand and a comment that this job only went to experienced men who were also trustworthy.

'You have to be able to read and write,' he said. 'There are books to be kept. We can't have someone like you doing this important job.'

I assumed that when he said 'someone like you' he meant an Irishman, and I was tempted to confront him, but the job was too important to me and I decided to try another approach.

'I've gone to college,' I said. 'I can read, write and count. I also have had experience keeping stock.' Then I played my Ace card.

'My name is Campbell. I'm sure you have heard that name around here.'

Grimes stared at me for a while, and then he mumbled that he would let me know later.

Later that day he told me the job was mine.

I felt a little guilty about using the Campbell name – suggesting a relationship with Andy Campbell – but I was determined to get the job and had no hesitation in using any weapon at my disposal.

I was also a little uneasy about being surrounded by enough explosives to blow me to kingdom come, but I told myself the new job was no more dangerous than working at the face, which had boxes of explosives all over the place. So, I abandoned the loco and began a new segment of my tunnel tiger career.

CHAPTER 9

A Resident Ghost?

I HAD ANOTHER REASON for wanting to get out of the tunnel, which was not related to my poor health. But I was too embarrassed to even mention it to anyone.

During my third week as a loco driver I came in on a Saturday afternoon to ferry workers in and out of the tunnel, but when I arrived there I discovered I was the only tunnel worker on the job. It was a very rare occurrence for three men not to show up for overtime but it happened in this instance.

I knew that there were a half a dozen skips that were full of gravel inside at the face. So I decided to go in and get them and dump them over the tip, so Griffin would see that I had something to show for my hours on the job.

I was in the best of humour because I thought I was going to have a very easy day and I was going to get paid double time for it as well. All I had to do was to take out the skips, tidy up a little at the face, and maybe change a few light bulbs.

But a strange thing happened to me when I had driven no more than two hundred yards into the tunnel – I was overcome with a feeling of dread, a panic attack that lasted for only a few minutes and then went away.

I was astounded by this attack, because I knew no basis for it. And as I thought about it I wondered what could have brought it on. As I proceeded deeper into the tunnel I felt a mild claustrophobia, which was something I had never experienced before, and I became embarrassed by my own nervousness. I thought that this was the way a small child would behave and not a nineteen-year-old man.

As I moved into the tunnel the panic attack faded, although

the mild sense of claustrophobia remained. I thought the cause of the claustrophobia was the fact that I was all alone in this place for the first time, and being alone was making me nervous.

My panic brought back a memory of an incident that occurred when I was fourteen, when my parents had gone away overnight to a wake and left me all alone in the big old hotel that was one hundred and fifty years old. I didn't sleep a wink that night and I heard sounds that may or may not have been real. I was a nervous wreck by the morning.

This is an experience I can still remember clearly. Anyway, I pulled myself together and got the skips hooked up to the loco and headed out. All was well until I was almost outside – in the same general location where I first felt panic – and then the panic returned.

Instead of making a run for it, I stopped the loco and shone a flashlight up and down the walls. But there was nothing unusual to be seen, although the sense of fear was as great as ever. I must have stayed in that same spot for half an hour before I headed for the friendly daylight outside the tunnel.

I had planned to go back inside the tunnel and work a few hours, but the more I thought of the panic attacks and the claustrophobia the less inclined I was to venture back in there again that day.

I did not consider myself a superstitious person up until that point. Indeed, I thought people who claimed to see ghosts under every bush were very immature people. But my experience in the tunnel had unsettled me. I came to the conclusion that this was an omen of some sort, a warning that something terrible was going to happen to me.

I may have inherited this belief in omens. My grandparents, Tim and Brigid Gallagher, who lived in Innisfree Island, off the Donegal Coast, were great believers in omens. They believed that banshees were sent to warn people of impending death, and sometimes a black dog would arrive at the door at midnight before a death. They were absolutely convinced of this.

Their beliefs must have seeped down into my subconscious, because here I was believing in omens, when all my life up to that point, I had thought these beliefs were examples of superstition.

In spite of my nervousness I hung around the outside tip all day, because I had no intention of losing the money I was to be paid by clocking out early and going back to the camp. When I got back to the camp that evening I said nothing about my experience to anyone because I was too embarrassed to talk about it. And I was a little nervous about going back to work the following day in case I had the same experience all over again.

However, I worked the following day without incident.

Several days later, while going into the tunnel in the middle of the week, I felt a jolt of fear as I passed the same spot. The panic attack was over in less than a minute, though it felt like it lasted for ages. At first I was determined not to be terrorised by this experience. I thought the best thing to do was to confront my sense of doom and stop at the very spot that was the focal point of my fears. Sometimes a week would pass and I would feel nothing, but then I would have the experience again and the terror would return.

The worst aspect of these incidents was that I was beginning to get preoccupied with this spot in the tunnel, and as time passed I wondered if I was not getting a little unbalanced and inflicting this situation on myself.

I was reluctant at first to talk to anyone, but as I thought about the experience I came to the conclusion that the fear was either an indication that something was about to happen to me in this spot, or something had already happened to someone else there – maybe a fatal accident – and somehow I was sensing it. Either that or I was slowly going mad.

I decided to explore my theory about an omen, so one evening while O'Donnell and I were down in the canteen having a beer after work I asked him casually if there ever had been any fatal accidents in the tunnel.

O'Donnell was a very perceptive person, and also a very superstitious one. He looked at me carefully and then, in true Irish fashion, answered my question by asking a question himself.

'Why are you asking?'

'I would just like to know.'

'But why are you asking at this time? There must be a reason. People don't ask questions like that without a reason.'

I didn't dare tell him the reason because he was so superstitious that if I told him he would have quit his job immediately. Eventually, after a great deal of prodding, he told me he had been on the job since the early months of the project and had never witnessed a fatal accident, but he had heard that a Donegalman had been killed on the job. He did not know at which site, however.

O'Donnell said the man had not checked a drill hole carefully and he had detonated a stick of unexploded gelignite when he began to drill. But O'Donnell claimed he did not know in which tunnel the accident occurred.

Somehow this did not make me less uneasy. My nervousness, combined with my chronic colds, eventually forced me try to change the job that I liked very much otherwise.

In the years that followed I did not hear of any fatal accidents in the tunnel, so I am at a loss to explain what was causing the panic attacks at that particular spot. Was it a resident ghost?

But if the threat was not real, I know the fear was very real, and to this day I wonder what caused it.

CHAPTER 10

Long Distance Kiddies

Among the thousands of workers on the construction projects in the Highlands were a small group of individuals who were known as 'long distance kiddies'. These were men who would work no more than a few months on any given project before getting an urge to pack up and move on to new pastures.

During their lifetime they wandered from Cornwall in the extreme south of Great Britain to the Shetland Islands in the extreme north

None of them ever gave a reasonable explanation why they were unable to settle down on any particular job for more than a few months. The only reason they ever gave was that it was time to move on.

Apart from their wanderlust, the long distance kiddies had very little else in common, except, perhaps, unlike the rest of the Irish workers, that they had not much interest in alcohol. They came in all sizes and in all ages, and their dispositions were as varied as their appearance.

I got to know two of these men during my stay in the Highlands. One was known as Wee Jimmy, a small, quiet, agreeable little man that everybody liked. The other, Big Johnny, was a hulking giant with an extremely bad disposition who did not have a friend in Dalcroy. I developed a relationship with both of these men.

Wee Jimmy's job at Bohesbic was to maintain the rail line outside the tunnel. I got to know him after I took over the job of powder monkey. The rail line ran for about half a mile from the mouth of the tunnel to the edge of a tip, where the rocks from the tunnel were dumped into a rock crusher that chewed them up into sand. This rail line was laid down on sand and gravel and it needed constant attention, because after the loco and skips ran over it several times it got twisted out of shape.

Jimmy was very conscientious about his work, always straightening out the rails or making sure that the rails were properly supported to carry the heavy skips. The little shack where I worked was very near the rail line.

I had very little to do on the powder monkey job and my entire activity was to sit around all day waiting for one of the tunnel tigers to come in for a supply of explosives.

On any given day I had no more than three customers and the rest of the time I sat inside the shack looking out the window at the spectacular Highland scenery.

After the brutally hard work I experienced inside the tunnel, and the health problems I had while driving the loco, I was thoroughly delighted with the new job because it paid a great deal of money and I had very little to do.

But after a few days of idleness, my active mind began to find this job a little too easy and I wondered what I could do to make it more interesting. Wee Jimmy provided a solution.

Wee Jimmy did not pay any attention to me during the first few days on the job. However, on the fifth day, there was a gentle knock on the door and, when I opened it, Jimmy was there. He asked me if he could come in. I was delighted to see him because, by this time, I was suffering from boredom and I was very glad of the company.

Jimmy had a habit of speaking his mind, so he came to the point of his visit very quickly.

'Nobody lasts very long on your job unless they are very old or very lazy. There is nothing to do and the days are very long and the first thing that you will be doing is talking to yourself. I know that in a wee while you will be trying to get another job, but in the meantime you should come out now and again and work on the rail line. Most of the others who had your job did that after a while. So, think about it and come out whenever you are ready.'

I was completely surprised at Jimmy's evaluation of my job. I considered myself very lucky to get this job as a powder monkey. Every evening, as I watched the scores of workers pouring out of the tunnel, filthy and exhausted, I thanked God I was no longer earning my pay in that way.

Every day I came to work with clean clothes on, and every evening I went back to the camp with my clothes as clean as they were in the morning. I didn't even get my hands dirty during the day. And best of all I had no supervisor, because Grimes never entered my shack, and I stayed away from the alcoholic store keeper's stockroom.

Three days after Jimmy visited me, I was out on the rail line working with him. After a week, I found myself staying out there most of the time, only going into the shack when someone came for gelignite.

From the first day I knew I had made the right move, because Jimmy was a born storyteller who had a repertoire of tales about his life as a long distance kiddie, and I never got tired listening to him. Jimmy was convinced that the present job he held was wonderful, because he was highly paid and he could easily handle the work.

'The first job I ever had was when I was sent off to Tyrone to work for a farmer, and me only ten years old. I did not want to go and work for a stranger for ten shillings a month, but my father and mother made me go, and they got all my earnings. I had to work sixteen hours a day, seven days a week, and I slept on hay in the barn at night. When the farmer offered me a job year round for ten pounds a year, they made me take it, and I never got to go home. When I was thirteen, I ran away to Scotland and got work picking potatoes for a Scotch farmer, and I got better money and better food. But the work was very hard for a wee fella like me. I never went back to Ireland after that.'

'You never saw your parents before they died?'

'They are not dead. They are still alive. He is eighty-eight. She is eighty-seven.'

I did not dare question Jimmy any further at this point because I knew from past experience that even though Jimmy could be very talkative about a wide variety of subjects, he was only talkative when he initiated the dialogue. Any attempt to cross-examine him on personal matters would be met with silence or an abrupt change of the subject. And anyone who persisted in spite of this would be permanently ignored by Jimmy.

However, I was very curious about whether this 65-year-old wanderer had ever contacted his parents in the 55 years that had passed since he left home, or if he ever planned to go and see them before they died.

Over the next few months he did give me information in dribs and drabs. I acquired other information from people who knew the family.

It seemed that Jimmy had never forgiven his parents for sending him off to work at such an early age, and he had no intention of ever going back to Ireland.

But he had been sending them money on a regular basis since he arrived in Scotland – money that he always mailed without an accompanying letter.

From time to time, the parents would reach out to him by sending messages and letters to Jimmy with tunnel tigers that were home on holidays. These tunnel tigers would deliver them whenever they ran across him. But the messages were met with silence.

As the weeks went by, I became very attached to Jimmy in spite of our age difference, and sometimes after work I would meet him in the canteen to have a few beers.

In time, I considered Jimmy a very close friend. I reinforced that friendship by being a good listener and never asking him too many personal questions. Jimmy seemed to return my friendship. However, one morning when I arrived at work I was surprised to discover Jimmy was not on the site, and as the day wore on I was worried that something had happened to him, because he was never sick and never missed a day's work.

I thought at first that he was just late, but as time passed I knew he was not coming in to work at all.

Since the camp was miles from the job site and there were no means of communication between the site and the hut where Jimmy had his bunk, I had to wait until quitting time before hurrying to the hut to find out what had become of him.

When I went into the hut and went up the long line of beds to Jimmy's, I discovered that a stranger occupied the bed. When I asked him where the old man was, he told me that Jimmy had packed his things and picked up his wages and left the camp. Jimmy had said it was time to move on – that he was a long distance kiddie and that he had a hunger for the open road.

I was stunned. I could not believe that he would leave like that without saying goodbye. I thought we were friends and that he would be considerate of my feelings.

I wondered when he had made the decision to leave. Was it an impulsive decision made in the middle of the night or had he planned it for days or weeks?

I was never to know. I ran into him the following year, but I kept him at arm's length and asked him no questions.

When he saw I was angry at him, he ignored me.

We rarely met after that and when we did we treated each other like strangers. I thought he was a strange little person that I could never get to know; he probably thought me equally strange for presuming that he owed me any explanation for his movements.

I first met Big Johnny when I was playing poker one evening with a group in my hut. We were playing penny poker and were having a good time without risking too much money. I was by far the youngest in the group and the only reason I had initially been allowed into the regular Wednesday night game was that the other regulars thought I would be an easy mark.

But after an hour's play, they quickly discovered that I was a shrewd and capable player who was nobody's fool. However, they allowed me to become a regular anyway, although I rarely lost much money. I enjoyed the poker games immensely and I looked forward to them. To me it was the social event of the week.

The night Big Johnny walked into the hut there were only five of us playing. Right away it was obvious that the other four knew

Johnny, and it was equally obvious that they did not welcome his presence.

The other four men greeted Johnny politely, but there was no warmth in their greeting, and no conversation with him at all.

Johnny was a very big man, well over six feet, and he was built like a heavyweight boxer. But it was his eyes that caught your attention first: they were small, cold, intelligent eyes that took in everything. That night as he watched the poker game, his eyes probed the players and the cards, observing every play, while watching the expressions on the faces of the players.

Johnny had a mannerism that I found very irritating. He always had several half crowns in his hands and he was forever playing with them, flipping them from hand to hand, or jingling them with his fingers. He also had the irritating habit of making comments about every hand played, most of them critical.

I think he was doing that deliberately to distract people, so they didn't keep their mind on the game. But I did not say anything to him because I did not want him to think that he was getting on my nerves.

Everybody knew that Johnny was a professional gambler. He was like the card shark that blew into town in the Old West. He was the hustler who tried to win a fortune at billiards or snooker.

Johnny lived for gambling: poker, horse and greyhound racing, Crown and Anchor, and even Pitch and Toss. He would not have worked as a labourer at all if he could avoid it, but nobody was allowed to occupy a bed in the camp if they were not part of the workforce, so Johnny was a reluctant tunnel tiger.

In spite of his size, Johnny was bone-lazy and only did enough work to avoid getting fired. But he was usually stuck with the least desirable jobs on the site, jobs no other man was willing to do, because the foremen disliked him and they made sure his life was not easy.

Johnny never got a job through influence – he always went directly to the office looking for work, and because of his hefty appearance he was usually hired.

Johnny stayed at any camp as long as there was action there. Whenever he could no longer get anyone to play poker with him or could not get a game of Crown and Anchor going he bailed out. The only money he valued was the money he won gambling, and he treated his pay packet from work with contempt, as if it were an insult to his intelligence, because he said he could make as much in one day gambling as he could in a month of hard labour. And he was right. When luck was going his way Johnny could win the equivalent of three months' wages in one night.

Of course, where there are winners there are losers, and anything that Johnny put into his pocket came out of the hard earned wages of other men – not that this concerned Johnny in the slightest.

Johnny was unpopular in the camps, not because he was a professional gambler, but because it was believed he was a cheat. None of this bothered Johnny, who did not care what people thought of him.

Johnny was also disliked because he was cold and mean. On top of that, he was considered a bully who only picked on those far weaker than himself. Still, very few were willing to challenge Johnny, because he had fists like sledge hammers, and even a coward can inflict serious damage if he is cornered and begins to lash out.

I was fascinated by Johnny the minute he walked into the poker game. I was also interested in the reaction of other players to him.

In spite of my youth I had been playing poker and had been around poker players for a long time, and I immediately recognised Johnny for what he was.

My experience at gambling began at an early age. My sister Annie, who was seventeen years older than me, had poker games once a week in one of the sitting rooms in Campbell's Hotel. For years I watched the players play the game.

I loved to watch the action and I very quickly learned the rules of the game. I also liked to move behind the players and noted what they did with the cards they were dealt. I loved a successful bluff.

One night, when I believed that I knew as much about the

game as the adults did, I asked to be included in the game. My request was greeted with laughter since I was only nine years old.

But when I persisted and showed them the fifteen shillings I had saved up, they said they would let me play one hand.

My sister thought it was all so cute, and the others patronised me as the hands were dealt. They did everything but pinch my cheek.

But I won that hand, and when I was allowed to play two more hands, I won those hands too. It was beginner's luck, and at this point the other players did not think I was cute anymore.

After that Annie evicted me from the game because she said if our mother came in and saw me playing poker we would all be sent to jail.

I was angry at the eviction, because by this time I was ahead ten shillings. I thought the adults were jealous of me because I was winning from them, and I thought they were embarrassed that someone so young could put one over on them.

However I soon found another game – a game that was played regularly at the local billiard hall – and within six months I had amassed the vast sum of one hundred pounds, a huge amount in those days.

But someone told my father about my activities. One night he came into the billiard hall and caught me in the middle of a game, and that was the end of my juvenile poker-playing career.

My father marched me out of the hall that night in front of a score of people, but not before he gave out to my fellow poker players, some of who were as old as seventy, for playing with me.

My father also confiscated the hundred pounds and never gave it back to me. It was not until I was sixteen years old that I was allowed to play with adults again, but by this time I was not nearly as lucky as I had been when I was nine.

During my earlier career I had learned to read the expressions on the faces of other players, and when I saw Johnny carefully scrutinise each player in turn I knew he was a wily opponent who was probably very skilled at the game.

When Johnny decided to join the game he did not ask anyone's permission. He just sat down and grabbed the cards and started to deal them. Since Johnny was a huge man with a bad temper nobody questioned his entry into the game.

I disliked Johnny's card playing style immediately and I could see why others would dislike it as well. When he was dealing the cards he gathered the discarded cards in front of him and always seemed to cover them with his huge paws, which led to the suspicion that he was picking cards from the heap, or shuffling the deck for the next round

Johnny was also frequently raising the stakes and trying to bully a player out of the game – which sometimes worked and sometimes didn't.

He got nowhere with me, however. When I called his bluff several times and demanded he keep his hands away from the discards, he first got angry with me and then treated me with a grudging respect.

The following evening, Johnny took me to one side and asked if I would like to be his partner. He said we could play as a team and share the winnings. I declined. I knew what Johnny's reputation was by then and I wanted no part of that ball and chain.

When I rejected him, he seemed to sense my reason and he said no-one need know about our arrangement. But I still said no because I wanted no part of this man or his schemes.

I wondered if Johnny had any idea about the bad reputation that he had and how this reputation had travelled all over Britain. Johnny could not go to a camp anywhere in Britain without being recognised, and once he was recognised the workers were very wary about gambling with him, although some still did.

I played a few more poker games with him. But in a few weeks it became difficult to get a game going because few people were willing to sit down with him, since Johnny would turn a game that people enjoyed playing into a contest that was stressful. I was the only one able to ignore his shenanigans.

When he no longer could get a game going Johnny set himself up running a Crown and Anchor game in the camp boiler room.

The game was very popular in all the camps in the north of Scotland. It has six symbols painted on squares. A die with all six symbols is thrown by the banker, and whoever has a bet on a square that matches the symbol on the top of the die when it stops rolling is the winner. The banker wins the money on the other five squares. There was no skill involved in playing this game – it all came down to luck.

The winner on any given square was paid two to one. In theory the banker can make a lot of money at this game, since he has only to pay out double on one square and he can pick up the money on the other five squares. But the game does not always work out to the banker's advantage.

There is such a thing as luck – good luck and bad luck – and luck plays a very important role in the game.

I was there the first night that Johnny opened his game and it did not go too well for him.

The problem started when one of the gamblers began to place big bets on the square with the anchor on it, and the anchor came up almost every time that this bet was placed.

The man was delighted at his good luck and the more cheerful he became, the angrier Johnny became.

I came to the conclusion that the man was in the middle of a lucky streak, so I followed him onto the anchor square with a bet and I started to win a considerable sum of money too. Big Johnny was furious.

Then many of the other gamblers started to pile on to the anchor and soon Johnny was in financial difficulties.

Johnny tried to end the streak by switching to a different set of dice. But this provoked a near riot because the gamblers suspected the new dice might be loaded and they threatened him with bodily harm if he did not continue to play with the same dice.

In the end, Johnny closed the game down because he said he was all out of money and could not take any more bets that night. The men did not believe this and some of them wanted to force him to empty out his pockets.

I did very well on the anchor streak because I was up a hundred pounds for the night, which was the equivalent of five weeks' wages. The gambler who began the streak was ahead four hundred pounds, and Johnny was out seven hundred pounds in all.

Johnny was by no means broke. Rumour had it that he had a bank account with several hundred thousand pounds in it and every pound was a prisoner. He liked to put money in the bank and never take it out again. So it was out of the question for him to go to the bank after he came to grief with the Crown and Anchor game.

He had an emergency fund of 200 pounds that he always kept in reserve, and he drew on that after he lost his bankroll. But since this was not enough to bank another game he began to look around for a partner. When a dozen gamblers turned him down, he asked me to be his partner.

Although I had turned him down when he wanted me to be his poker partner, I accepted when he asked me to go in with him on the Crown and Anchor bank.

I thought I had a very good reason for this. First of all the only reason he wanted me in the poker game was to collaborate in some scam he had in mind to cheat the other players. I wanted no part of that.

But the Crown and Anchor game involved pure luck, and the odds favoured the bank, so I thought it a good risk, even though I did not trust Johnny and was determined to keep an eye on him.

I also liked the idea of being a banker – the title had a nice ring to it and it fed my ego. Whatever reservations I had about Johnny were overshadowed by the lure of fame and fortune.

The first night of our partnership went well until it came time to divide the spoils. Johnny had been taking in the money all evening and when he counted it after the game was over, he said there was only two hundred pounds profit: one hundred pounds each. I was furious because I knew the profits were much greater than that.

I told him we were finished, and when he saw I was serious,

he came up with another fifty pounds, saying he was giving me some of his own money. I told him he was a liar.

I only agreed to be his partner the following night when he agreed that I would take in the money the first half of the night and he would be the banker the second half. I knew he only agreed to this arrangement because he still needed me. If he didn't, he would have told me to get out of his sight. I also knew that if he had another good night, I would be fired and, once again, he would have all the action to himself.

Things went very well for the first two hours when I was the banker, and I had three hundred pounds profit in my pocket when it came time for Johnny to take over. He wanted me to hand over the money so 'he could take care of it', but I told him I was well able to take care of the money, and I held on to it. This infuriated him but he did not want a row in public so he said nothing.

The second half of the night was going even better, and Johnny soon had a wad of money that he had difficulty getting in and out of his pocket.

Johnny was delighted at his good luck. Having a big, wide mean streak in him, he could not prevent himself from smirking at the losers and congratulating himself with each successful throw of the dice. Then, the lights suddenly went out and there was pandemonium in the pitch-dark room and a great deal of shouting and cursing.

When the lights first went out an instinct told me that this blackout was no accident. I instantly took cover under the heavy table.

My instinct was correct: under cover of darkness one or more men jumped Johnny and relieved him of his wad. I don't know if they intended to do the same to me, but they did not get the chance since I was so quick to duck for cover.

When the lights went on the place quickly emptied. Johnny went outside trying to find the thieves, yelling about being robbed and looking for the men who had robbed him.

While he was out there and the boiler room was empty, I hid the money I had kept away from the robbers deep in a cupboard in a corner of the room and waited for Johnny to come back.

When he returned I began to complain about being hit over the head with a stone. I also told him my money had been stolen. Smart as he was, Johnny had no reason not to believe me, and he was sure I too had been a victim.

The following morning Johnny picked up his work papers and his wages and went on the road again.

I retrieved the money from the boiler room later without any qualms of conscience, because I believed I was only getting back the money that had been stolen from me the first night.

I never ran the bank again because I thought it too dangerous. But I have always remembered those two nights as being very exciting – indeed they were the most fun I have ever had at gambling, right up to the moment when the lights went out.

CHAPTER 11

Predators

BIG JOHNNY WAS A man who did not like to play by the rules: he was a predator who was always on the lookout for a way to take advantage of any given situation. Rules were for other people.

However, there were other predators at Dalcroy who did not want to obey rules either, especially the laws that prohibited the killing of game in the hills and lochs around the camp. These men believed that since they were not being given the good fresh food they thought they were entitled to they could help themselves to the free food that was available in the countryside, food such as venison, salmon and grouse.

The venison was on the hoof, of course – the deer had to be brought down with rifle shots, slaughtered, and dismembered out in the countryside. The grouse also had to be shot, and the salmon had to be lifted from the rivers.

Technically, this was not free food. The salmon, grouse and deer were on the estates of major landowners and were considered the property of the landlords. These landlords hired bailiffs to watch over their game, but the wily poachers were adept at keeping an eye out for bailiffs as they helped themselves to the fresh meat.

Some of the men involved in the poaching were Irish; others were Scots. The men who engaged in this activity were not hunters in the commonly accepted view of the activity. They did not acquire licences and they were not engaged in the activity mainly as a sport. They were poachers who were out to acquire free food, and the thrills came from doing something illegal and the ability to fill their bellies with contraband.

These men were obviously predators, but unlike Big Johnny who victimised his fellow workers, these men made victims of the landlords, and the unfortunate animals who became their prey.

Preparation for the foundations of Errochty Dam, which was to be built from the photo point towards the two masts, known as blondins. Blondins were overhead cableways almost half a mile long stretching over the valley; they were used throughout the whole construction site to transport material and pour concrete wherever required for the dam construction.

A M Carmichael, Sr, 1950; collection of A M Carmichael, Jr.

Left: Excavating the hillsides for the construction of the Errochty Dam was done by blasting and use of heavy construction equipment.
A M Carmichael, Sr, 1950; collection of A M Carmichael, Jr.

Below: A six-mile tunnel was contructed from the Errochty Dam site to the shores of Loch Tummel, where the largely underground Errochty Power Station was being built. 1,000 feet (305 m) into the hillside, the main tunnel split into three smaller separate tunnels which diverted water to three turbo generators in the power station. The three tunnel entrances can be seen here.
A M Carmichael, Sr, 1950; collection of A M Carmichael, Jr.

Foundation work for the power station. Much of the 450,000 tons of rock excavated from the principal tunnel was used for aggregates in concrete for the dam, and base material for access roads. Errochty Power Station was also aesthetically faced with a mixture of extracted schist and quartzite. Some 1,525,000 lbs (3,362 tons) of explosives were used in the tunnel.

A M Carmichael, Sr, 1950; collection of A M Carmichael, Jr.

Erection of the steel structure for Errochty Power Station.
A M Carmichael, Sr, 1950; collection of A M Carmichael, Jr.

The power station – which would handle 1,000 million gallons of water per day from the tunnels – takes shape. In March 1952 the Tunnel Tigers achieved a world record on the vertical shaft – 600 ft (183 m) – which connected to the main tunnel.

A M Carmichael, Sr, 1950; collection of A M Carmichael, Jr.

Construction of ancillary aqueducts and tunnels was also part of the massive project. In addition to the principal six-mile tunnel, a further 12 miles of tunnel were built to collect water from the rivers Garry, Bruar and Edendon to feed Loch Errochty. This was possible because rainfall in this catchment area averaged 55 inches (140 cms) per annum.

A M Carmichael, Sr, 1950; collection of A M Carmichael, Jr.

A M Carmichael Ltd operated a huge fleet of lorries and construction equipment and a good number of vehicles were sourced from the British War Department, these having seen service in WWII. This Chevrolet CMP tipper is such an example, and in the third photograph a Bedford QL Command Truck can be seen in the foreground.

A M Carmichael, Sr, 1950; collection of A M Carmichael, Jr.

Errochty Dam as it can be seen today, if one can find its secluded location in the Perthshire hills. The dam has 28 buttresses, is 1,400 ft (430 m) long and 120 ft (37 m) high. 250,000 cubic yards of concrete were used in its construction.

A M Carmichael, Jr, 2003; Collection of A M Carmichael, Jr.

All of the men involved argued that there was nothing wrong with helping themselves to game, because they believed the landlords had no right to lay claim to the game in the first place.

This was a subject that the Scots and the Irish agreed on – one of the few things that they agreed on – that landlords had no right to restrict access to game, and that ownership of the land did not mean the landlords had a right to everything that walked on it. These poachers were happy when one of their hunting expeditions came to a successful conclusion, and they were inclined to boast to everyone in the camp about the latest kill.

There is little doubt that the officials in the Carmichael organisation knew all about the poaching but chose to do nothing about it. Cracking down on poachers would win no popularity contests for Carmichael and the best thing to do was to turn a blind eye to the poachers, and leave it to the landlord's bailiffs to go after the predators.

The salmon were taken out of the river after being electrocuted by a device the poachers had invented for the purpose. A wire attached to a metal grid was dropped into the river near a major salmon run. When salmon were seen in the vicinity of the grid, the power was turned on and the fish electrocuted.

The killing of the salmon never took place near a village or near the camp at Dalcroy, because there would have been too many witnesses around. Instead the salmon were lifted in the Tummel River near the construction site, where there were few people, and an ample supply of electric power to construct the deadly salmon trap.

The salmon were consumed at the construction site. They were grilled over a coke fire and seasoned with butter, pepper and salt. To ensure that no one squealed to the bailiffs, the fish were shared with the workers who were assigned to the area, thus making them accessories to the crime.

During the years I was at Dalcroy I do not remember an incident in which any of the men were caught poaching salmon, so the bribes seemed to have been effective.

Killing the deer was more difficult, for more reasons than one.

First of all the deer had to be killed on the open hillside, and as a result the hunters were in danger of being seen by the bailiffs. Then there was the noise of the rifle shots that also attracted a great deal of attention. The logistics of getting the kill back to the construction site over very rough terrain was another problem. This was overcome by butchering the deer on the hillside and taking it down in sections.

On one occasion, two hunters named Michael Campbell and Jim Forsythe were out hunting near Garve, Ross-shire, when they saw they were being watched by two armed bailiffs. Campbell fired a shot high over their head and this frightened the bailiffs, who thought that killers were trying to assassinate them. The bailiffs skedaddled.

The fresh venison was also shared with co-workers, and so were the grouse As a result, many of these workers were living a very grand lifestyle, eating gourmet food, all of it fresh, all of it free.

Once these men had acquired a taste for this exotic food, it was next to impossible to tolerate the food served in the mess hall.

A few of these hunters told horror stories about the food in the mess hall to justify their poaching activities out in the mountains.

Two of the favourite tales involved the huge cat population that at one time plagued the Dalcroy camp.

According to one of these stories, a cat fell into one of the huge vats of porridge that was cooked every morning and served to the workers for their breakfast. The cat was dragged out of the porridge and his carcass thrown out the back of the mess hall, and the porridge was then served as if nothing had happened.

Shortly after that the cooks at the mess hall announced that they were serving rabbit in order to vary the diet from the tasteless mutton that was served almost every night. They said the rabbit would taste a lot better than the mutton.

No sooner had the rabbit made its appearance on the menu, than the cats began to disappear, and the rumour got out that the men were being served fresh cat instead of fresh rabbit.

These stories were a good excuse for the hunters to stay away from the mess hall, and they provided them with a justification for helping themselves to the game out in the countryside.

However, the majority of the men ignored the rumours and went right on eating the food in the mess hall anyway.

Many of the hunters needed no excuse for going after game, because this was an activity that was very popular in both Donegal and the Highlands of Scotland.

I remember well the thrill of going out on the Dungloe River in Donegal in the dead of night with fishing net in hand, and in a remote area tossing the net across the river and then waiting until trout got entangled in it.

Catching trout was a major thrill, but not nearly as thrilling as the thought that the bailiffs could come at any minute and drag you into court for the illegal use of a net to catch fish.

It was that excitement that added the spice to the adventure and made it so enjoyable.

The poaching of salmon and the killing of deer was common in many communities throughout the north. But the men in the camps who were involved in it took a special pleasure in bringing into their diet the type of food that could be only dreamed of by the men who were subsisting on bad tasting porridge and rancid mutton. The fact that this poaching made them very popular with the rest of the construction workers only added spice to the adventure.

The odd thing about the poaching was that it brought some Scots and Irish together in a way no other activity could. While they were out together stealing from the landlord there was no word about religious or ethnic differences. Instead they acted as if they were old friends.

Although I was not directly involved in the poaching, I did not see any harm in the men helping themselves to free food at the landlord's expense. Vast stretches of the Highlands were owned by absentee landlords who lived in London, and these estates had been seized from the ancestors of the Highlanders without any compensation.

So, helping oneself to a little game was a minor crime in comparison to helping oneself to vast estates.

CHAPTER 12

Rebels

WHEN I FIRST ARRIVED in the Highlands relations between the Irish and the Scots were very cool – to put it mildly. Then for a brief time in the early Fifties the Irish Republican Army (IRA) began to revive their campaign to rid Northern Ireland of the British, and relations went from bad to worse. Somehow, the 99% of the Irish in Scotland who were not in the IRA were being held responsible for the activities of those who were.

The major activity of the IRA at that time was a campaign to acquire armaments, and in order to achieve this they raided British army barracks for guns, and construction sites for explosives. The campaign fizzled out after a short time and was not revived on a major scale until the Seventies. Although the IRA did not make many spectacular raids on Britain during the Fifties, these raids nevertheless generated a lot of publicity in the newspapers and revived the image of the Irishman as a dangerous anarchist.

Even though there were probably no more than a hundred volunteers in the IRA during the Fifties, many of the Catholic Irish in Britain were branded with the violent image of the IRA, as if somehow they were involved with the rebels and were aiding and abetting their activities.

There was no way the Irish living in Britain could combat this image, because the British had a bad attitude towards the Irish to begin with and the activities of the IRA just reinforced it.

The attitude of the men in Dalcroy Camp towards the rebels was ambiguous. Although none of the workers that I knew were directly or indirectly involved with the IRA, they had a fascination with the IRA gall in taking on the British Army. During this time I never heard anyone say the IRA was wrong or foolish, or that what it was doing was against the laws of God and man.

There were a couple of reasons why the men in the camp had

this ambiguous attitude towards the IRA, even though the IRA was causing problems for the Irish living and working in Britain.

First of all most of the men in the camp were from Donegal, a county in the north west of Ireland. Since Donegal was one of the three Ulster counties in the Irish republic, the people of this area were inclined to see the other six Ulster counties that were part of the United Kingdom as 'occupied territory.'

Most of the Donegalmen believed that the British had no right to be holding on to this part of Ireland, so there was some sympathy for the IRA trying to run them into the Irish Sea. But the men thought that the way the IRA were going about it – a tiny group of men fighting with stolen weapons – was guaranteed to fail.

The Irish Catholic workers at Dalcroy were also hostile to the Ulster Protestant workers who refused to have anything to do with Catholics. Since the IRA was rocking the boat for the Protestants back in the Six Counties, the Irish Catholics felt the Protestants richly deserved it.

Then, there was the anti-Irish prejudice that the Irish were experiencing in Scotland, which was resented by the Irish workers. This prejudice led to a malicious enjoyment of the rage expressed by the Scots over IRA attacks.

And yet, if the Irish workers were asked to name the wrongs that they had suffered in Dalcroy – apart from the barely concealed hostility – they would have had a hard time coming up with a credible list. The reality was that the Irish in Dalcroy were among the best-paid labourers in Britain, who earned more money and were able to save more money than any of their British counterparts.

The working conditions were, of course, tough, but the workers did not even have to put up with tyrannical British bosses, because the bosses in Dalcroy were mainly Irish. If workers were abused occasionally, it was by their own kind.

Anyway, even the Irish bosses were never known to abuse Irish workers on a continuous basis, because these bosses were also mainly from Donegal. If a worker was abused, there would always be a day of reckoning back in the home parish.

The truth was that the Irish had a type of home rule in Dalcroy as there was only a single policeman in the camp, and he rarely left his tiny office. Troublemakers were evicted from the camp by being fired by the Irish foremen. The camp was in fact a self-regulating independent community that needed no administration by the British Government. This independence was reflected in a couple of incidents that occurred in the early Fifties when I was there.

The incidents involved two major events in British history: the death of King George VI and the coronation of Queen Elizabeth II.

When the death of King George VI was announced the Carmichael Company decided to ask all workers to observe two minutes of silence the following day. All work was to cease; all machines were to be shut down.

When the time came for the two minute silence, Michael Campbell, the son of Andy Campbell, who was a mechanic in the Black Gang, decided to make a gesture of his own – a gesture that illustrated how fed up he was with the way he was treated by British members of the Black Gang.

Michael claimed later that he had suffered a daily barrage of abuse in the Black Gang because he was an Irish Catholic, and he was subjected to this even though he was the son of the superintendent. So, Michael had been building up a head of steam for some time.

Campbell's gesture was to turn on the air horn used to mark quitting time, and he sounded it for the entire two minutes. He said later that he had not planned the gesture and it was in reaction to some slight he had received earlier in the day which was the last straw as far as he was concerned.

The gesture created an uproar among the Scots and Protestant Irish, which was understandable.

There was little doubt that Campbell would have been fired were it not that he was the boss's son.

Michael was unrepentant, however, and he said later that his father never criticised him for the incident because he knew that those who were slighting Michael were also slighting him.

'He knew what I was going through in the Black Gang, and he

knew that I was leaving to work in America because of what I went through in the Highlands,' Michael said.

The second incident occurred on the day of the Queen's Coronation. On that day, the sole policeman in the camp decided that a Union Jack would be flown at the entrance to the camp to mark the event. Red, white and blue bunting should be strung up to lend a festive air to the community, he said

When word got out that the policeman planned to fly the British flag the policeman was told by some of the men that this would not be a good idea because the Irish in the camp might not like it. He thought they were joking and he said he was going ahead with the flag-raising anyway.

On the morning of the event, the policeman came out of his office with the Union Jack and was met by a crowd of Irishmen, numbering in the hundreds, who stared silently at him as he headed for the flagpole. Although no threats were uttered the policeman lost his nerve and retreated to his office and called the Pitlochry barracks for reinforcements. He said later that he was determined that a crowd of unruly foreigners was not going to prevent a British flag being unfurled on British soil.

But when he explained the situation to the police headquarters in Pitlochry he was told there were not enough police in Pitlochry to handle that many Irishmen if they decided they didn't want the flag unfurled.

When the lone policeman at Dalcroy asked his supervisors in Pitlochry what he should do about the flag, he was told that this was a decision he had to make himself, and he had to accept responsibility for that decision. In other words, he was on his own.

The flag did not go up that day and nothing more was said about the incident; but I am sure the Scots in the camp were given one more reason for disliking the Irish.

In a way the flag incident was an example of how the Scots and the Irish had developed such a poor relationship with one another.

The Irish were furious at the slights they experienced in Scotland and they rarely missed an opportunity to let the Scots know how they felt about it. The Scots resented the hostility of the Irish, and so they continued to express contempt for them. They were also furious that the Irish had the nerve to confront them on their own soil. As far as the Scots were concerned, the Irish should have been grateful that they had good jobs.

*

My own attitude towards the IRA and the British establishment was also ambiguous. I recognised that I was earning an excellent wage in Britain, and I had every reason to be grateful for this. But I, too, felt the snubs, and this diluted any gratitude I may have felt.

Anyway, I certainly felt no need to salute the royals.

As far as my attitude towards the IRA was concerned, family links with the organisation went back to 1916. Three of my father's cousins, Joe Sweeney, Peader O'Donnell and Hugh Doherty, had been leaders of the IRA back in the Irish War of Independence.

They had been with the leaders of the IRA, Pearse and Collins, in Dublin when the insurrection began. They continued with the hostilities until they won twenty-six counties from the British, leaving the disputed six under British rule.

Joe Sweeney and Peader O'Donnell split with each other over a ceasefire treaty that allowed the British to remain in Ireland. They fought each other in the Civil War that erupted after the signing of the treaty.

Sweeney, who sided with the pro-treaty faction and later became chief of staff in the Irish army, backed the winning side; O'Donnell, who had thrown in his lot with De Valera, the IRA leader who opposed the treaty brokered by Collins, was on the losing side. O'Donnell later became a well-known writer and socialist who fought for better working conditions for the immigrant Irish labourers working in Scotland.

When I was growing up both men visited our house but at different times. My father was a supporter of Fine Gael, the political party Sweeney belonged to; my mother liked Peader O'Donnell because he had once been her schoolteacher. So I heard both sides of the issue, including the one thing both Sweeney and O'Donnell agreed on – that the British had no business remaining in Ireland.

My own attitude towards the Anglo-Irish and the British became a little more complex as I grew older.

By the Fifties, my sister Rose had married a Scotsman, Jimmy Stamper, who had served in the British Army in Northern Ireland during the late Forties.

My sister Catherine had married Arthur Bardon, a Catholic with Irish Protestant and English roots. Since I was very fond of both men, I did not automatically reject either Scotsmen or Irish Protestants. I only rejected the ones who gave me reason to be annoyed at them.

But in my heart I did not believe that the British should be in Ireland and I, too, deeply resented the hostility I was exposed to in Scotland. I was inclined to think that somebody had to let the British know how the Irish felt about them.

Even so, I was embarrassed by the flag incident at Dalcroy on the Queen's Coronation Day. After all, this was the United Kingdom and I thought that to prevent Britain's flag from being flown in Britain was comparable to the British refusal to allow the Irish Tricolour to be flown anywhere in Northern Ireland. I had always resented the fact that the Irish Tricolour was banned in Derry, Strabane, and Newry, towns with huge nationalist majorities, because these were essentially Irish towns with a loyalty to the rest of Ireland, and I believed that the nationalists in those areas should have been allowed to display the flag if they wished.

Dalcroy, on the other hand, was in British territory, and the Irish workers were in fact guests. So I thought they could have displayed a little more common sense and not made an issue of the flag. However, I said nothing at the time because most of the rest of the men were intoxicated with the power they had acquired by preventing the raising of the Union Jack, and I had no intention of trying to reason with a group that was riding high on a tide of emotion.

CHAPTER 13

A Workers' Democracy

ALTHOUGH THERE WAS A class system in place elsewhere in Britain, the camp at Dalcroy was devoid of all social ranking, and even the supervisors enjoyed little esteem from their position.

It made no difference to most workers what social position a worker had prior to his arrival in the camp. The minute he arrived there he had the same status as every other worker in the place.

There were a number of men in the camp who had come from families who thought they were a notch up on the social scale, but these men were now working with a pick and shovel like any other labourer. For some of these, the adjustment to camp life was very difficult to handle.

Although I had come from a family that owned a business, I had no problem at all handling the transition, because the camp did not represent a step down for me. On the contrary, the work, the pay, and the working hours were far superior to the situation I had left behind. Indeed, I felt I had been released from slave labour and moved up in the world.

But many of the other workers who had a similar background had great difficulty with the democracy of the camp, especially those who had been proud of their social position back in their home country and now found themselves working, sleeping, and dining with men that they once thought were their social inferiors.

The maladjustment was expressed in many ways and had different degrees of severity.

Stefan, a Polish born worker, was typical of this breed who could never adjust to life on the camp, because he lived almost entirely in the past and was never prepared to face the reality of the present.

Stefan rarely engaged in a conversation with anyone, but when he did it was always about his life in pre-war Poland and never about the present or his future in Scotland.

Stefan was convinced that he was an aristocrat living among peasants and he seemed unhappy every moment of his life in Dalcroy. He did indeed have the air of an aristocrat about him and he had the type of personality that immediately caught your attention.

Stefan first joined the German Army during World War II as part of a Polish brigade fighting for the Nazis. He had been captured by the British and later interned in a POW camp in the Highlands until 1948.

Stefan had been a high-ranking officer in the German Army and when he arrived at Dalcroy he had worked for a brief time as a labourer. Andy Campbell had spotted leadership qualities in him and put him in charge of one of the shifts in the tunnel.

I knew Stefan because he would come to me for explosives, and I found him a polite, refined individual who seemed to be out of place in a construction tunnel. Stefan had been an instructor in a Polish university before the war, and he had the air of an academic who was dressed up in filthy construction clothes for the day.

The Polish foreman did not seem to have any friends. There were other Polish labourers on the site but they would have little to do with him because he had switched to the German side after Poland had been defeated.

There were a number of other Poles in the camp who had also fought with the Nazis, but for some reason these Poles never associated with Stefan either. Some of these Poles suggested Stefan had been involved with the concentration camps and committed unspeakable crimes against Jews and others viewed by the Germans as enemies.

There were many Germans in the camp who were also ex-POWs, but he was not particularly close to them either, although I did see him speak to them occasionally.

He was however an excellent foreman who could get the work done and he had the respect of those who worked under him.

Stefan could never return to Poland because of his war record. He seemed homeless and isolated, but he appeared to like it that way. He never extended the hand of friendship to anyone, and he rebuffed anyone who reached out to him.

Stefan left the camp every Saturday afternoon and did not return until Monday morning. A German acquaintance of his said he had a house rented outside Pitlochry and that he spent his weekends there listening to classical music and walking in the mountains. Every Sunday afternoon a young woman visited him.

Stefan was at Dalcroy when I first arrived there, and he was there when I left.

Sometimes, I wonder what became of him.

*

A Donegalman whom I will call John T. was even more at odds with the world than Stefan was.

John had been a successful businessman who had blown away a fortune on horses and whiskey. He arrived at Dalcroy wasted both physically and mentally.

John was given a pick and shovel job when he arrived but he could not handle the hard work and Andy Campbell gave him a job in the stockroom. He could not cope with that either, so he wound up doing the lowest job on the site: making tea for the workmen and running errands for the bosses.

John spent his evenings in the canteen drinking beer, and he came into work every day with a hangover. This eventually put him at odds with the men on the job who had felt sorry for him.

John never drank on the job, because he knew this would not be tolerated. Drinking on the job was not tolerated by anyone, because this was a very dangerous construction site and a worker had to have his wits about him at all times.

If a worker arrived at work drunk, or drank during the day, his co-workers would turn him in right away. A boss who caught a worker drinking would fire him on the spot and escort him off the site.

At first John managed to confine his drinking to the canteen in the evenings, but after a month he began to drink before he went to work. Soon he began to show signs of inebriation early in the morning.

Apart from his drinking, John's biggest problem was that he seemed unable to face the new reality he lived in.

All his life he had been looked up to because his family was wealthy and he had inherited a great deal of wealth from his parents. He seemed to expect that the working men at Dalcroy would look up to him because of his family background, and he had great difficulty accepting that he was being treated like just another labourer.

But the tunnel tigers were not being unkind when they failed to humour John. All of them worked very hard for their money and they had no time to play make believe with someone who thought he was better than they were.

To their credit, they never let him know this. They just listened politely when he began to ramble on about the days when he had a staff of his own and more money than he knew what to do with.

Most felt sorry for him.

One morning the word came up from the office that John had to go. The men made a collection and he was handed a sum that would keep him drinking for a week. Then he was put on a bus for Perth and that was the end of his tunnel tiger career.

I ran into John frequently in Perth during the next few years. He never worked again and hung out in a bar that catered to the Irish, sitting in a corner waiting for someone to buy him a drink or a meal. He seemed to survive.

John had a wife and children at home, but he never went home to see them. Booze had made him an empty hulk that cared about nothing but where the next drink was coming from.

*

Joe had also been an alcoholic businessman prior to his arrival in Dalcroy. While he was very unhappy at his new situation, he kept his drinking under control and was always cold sober on the job.

In spite of the fact that he was twice my age, he and I hit it off and we became close friends.

Joe was into total denial about the reasons for his downfall and held a score of grudges against people whom he believed contributed to his situation. Prominent among these were people who had loaned him money and had been putting pressure on him to pay it back. They were all 'scum of the earth' as far as he was concerned.

Joe managed to get himself into one of the huts where there were a score of cubicles, each with two beds. He was very proud of the privacy he enjoyed even if he had to share the cubicle with a Scots mechanic who didn't like him.

I envied his sleeping accommodations, since I lived in a hut with forty beds – all lined up close together.

One day he told me the Scots mechanic was going home for a week's holiday and that I was welcome to have the Scotsman's bed while he was away. I asked him if he had the mechanic's permission, and he said he had – which was a lie.

I enjoyed my week out of the noisy hut immensely, and I looked forward to the next time the Scotsman went on holiday. But a few nights after the Scotsman's return, Joe arrived in my hut carrying his belongings and took up residence in an empty bed.

I went over to him and asked him why he was in the hut. He told me that he had been kicked out of his cubicle and that it was all my fault.

'They put me out because Frazier complained that I had let an Irishman use his bed when he was away. He demanded a new mattress, and he demanded that I be evicted for breaking the rules.

'Nobody is supposed to be in those huts except tradesmen, and an exception was made for me. They said you broke the rules when you slept in the cubicle and I let you. This is entirely your fault.'

I didn't bother pointing out to him that he was the one who created the situation by inviting me into the hut without Frazier's permission. Joe had such a victim complex that he would argue black was white. You could not win an argument with him.

But the bit about Frazier demanding a new mattress because I had slept in it annoyed me.

'Why did he ask for a new mattress?'

'He said he was not going to sleep on any mattress that a low class Irishman had been sleeping in,' said Joe. 'I told him you were not low class... that you came from a good family like mine ... that your parents owned an hotel. But he said he didn't give a damn if your parents were Irish royalty. He was not sleeping in any bed that an Irishman had been sleeping in.'

'And what did you say?' I said.

'I said nothing.'

I did not say anything further about the issue to Joe, but I was furious at Frazier for talking about me like that.

*

I thought that Frazier was very foolish to go around talking about the dirty Irish and he was really looking for trouble. There were a thousand Irish in the camp and only twenty Scots, so Frazier was obviously not thinking too clearly. Common sense would have dictated that I ignore the slur, but I exhibited little common sense.

The remark about the mattress really annoyed me, much more than any other slur directed at me since I arrived in the Highlands. It was a sort of last straw that I found totally unacceptable.

I decided that I was not going to let this man get away with it. I would find a way to get even.

I did not tell Joe about what I was thinking, since he was useless at confrontation and would appease or run away rather than stand his ground.

But I knew a few other men in the camp who did not take insults from anyone. I approached a couple of them about Frazier and his prejudices.

Both these men were big, tough-looking Irishmen. When I approached Frazier in the canteen that night with them by my side I saw he did not like the expression on their faces.

'I hear you think Irishmen are dirty,' I said. 'You said you would never sleep on any mattress that one of them had slept on. How dirty do you think we are?'

It was obvious Frazier was panic stricken, and so were the men who were with him, because they got up and left the canteen.

However, before he had a chance to say anything, a Scotsman who was one of Andy Campbell's assistants came over and told us that anyone who started a fight would be escorted out of the camp that night.

'You know the rules. You fight; you leave.'

We turned away – Frazier was not worth the loss of a job – but before I left, Danny Campbell, one of the men who was with me, told Frazier that this was not over – that there would be another day. We then walked out of the canteen and left Frazier sitting at the table alone, deserted by his friends and isolated from the rest of the men in the canteen.

Frazier left the camp the following morning and was not seen in the Highland camps again.

I had mixed emotions about the outcome of the confrontation, but I was glad that I had made Frazier fear us, even if he did not respect us.

But Joe was furious at me for staging the confrontation with Frazier.

'You are picking up all the bad habits of the hard men around here. What do you think this is... the Wild West? People like you give the Irish a bad name. People like you are always badmouthing the Scots, but when they say something about you you want to start World War Three. You will get us into trouble. You will get us fired.'

Still, I had no regrets. And I had no feelings of guilt about chasing Frazier out of the camp. But I knew Joe had made a good point – the Irish exhibited a great deal of prejudice themselves, and were often as guilty as the Scots.

Joe kept his distance from me for a time after that because he was afraid I would get him into trouble. After a time he started to come around again, since he had no-one else to turn to. Joe was really harmless and he was good company when he was not boasting about his pedigree.

*

The occasional ethnic friction between the Irish and the Scots was the only thing that marred the democracy that otherwise governed relationships in these camps.

Men were respected for their hard work and even their thriftiness. A family man who saved every penny so he could give his children a good life back in Ireland was esteemed.

Drunks or gamblers of any nationality were not respected. Those who were lazy were barely tolerated. Nobody complained if these men were fired. Given the work ethic and the lack of pretence, those who arrived with big ideas about their own importance had to adjust quickly – or leave.

CHAPTER 14

Conflicting Loyalties

THE CO-OPERATION AND SHARED work ethic that existed between men of many nationalities and many different religions while on the job did not always carry over once the day's work had ended. Driving tunnels and building dams was very dangerous work, and each man instinctively knew that no issue involving race or religion had any place on the job, but once the day's work was ended the divisions between the various ethnic and religious groups in the camp surfaced.

The Ulster Protestant workers had no conflict with the Irish Catholic workers out on the construction site, but once the day's work was over, the Irish Catholics headed for the camp at Dalcroy, and the Protestants headed for a different camp, because they did not want to share living accommodations with the Catholics.

The Scots members of the Black Gang had their separate accommodations also, and Irish Catholics were not welcome in their area.

The Scotsmen from the Hebrides, and the German and Polish workers, were usually integrated with the Catholic Irish and there was rarely any conflict between these groups on or off the job.

Open hostility between the Irish and the Scots and Ulster Protestants was very rare, because Carmichael would not have tolerated it. The company treated everyone fairly and company officials would not accept open bigotry from anyone, no matter who they were.

The ethnic and religious tensions were reflected not in open hostility but in the unwillingness of any party to the conflict to treat one another with respect or to socialise after the day's work was over. They instead ignored one another, or engaged in snippy remarks that were a reflection of a barely concealed prejudice. No one ethnic group had a monopoly on this type of behaviour, however.

As the months passed, I adapted to the conditions of living and working in the Highlands. I looked, dressed, and acted like any of the other multi-national tunnel tigers who soldiered along with me in those mountains.

There was one aspect of life in the Highlands that I never got used to – this undertone of tension between the Irish and the Scots and Ulster Protestants who worked on the project. I was annoyed at the fact that the Ulster Protestants would not live in the same camp as the Catholic Irish, and I resented the cold formality that Scotsmen exhibited in any interaction with me on or off the job.

I had arrived in Scotland with little notion of the extent of the negative attitude that was directed at the Irish, but had instead come with the idea that I was coming over to the ancient homeland of the Clan Campbell. My father always talked about our Scottish roots, not only because of the Campbell name, but also because of his mother's Mac Sweeney name, a name with Scottish roots that was also in our family.

I quickly learned, however, that my Scots roots did not mean very much to many of the people I met in Scotland. Instead their negative attitude was based on the fact that I was a Catholic and was born in Ireland.

I had never been directly exposed to ethnic animosity while living in Donegal, because north west Donegal was mainly Catholic, and the Protestants who lived there never expressed any hostility towards me. I was well aware however of the animosity of the Orangemen who lived over the border.

So, the first taste of prejudice came as quite a surprise to me, and as I became aware of the extent of the prejudice my resentment deepened.

If I had been a little more perceptive at the time and a little more honest with myself I would have realised that there was hardly anything unique about the prejudice of the Scots.

After all, I had been exposed to the bitterness expressed by many Irish Catholics towards Ulster Protestants, and to the complete opposition by Catholic parents in north west Donegal in the

1940s to any intermarriage with local Protestants, Jews or people of colour.

I had also been exposed to the prejudice against tinkers (itinerant tinsmiths) all over Ireland, who were viewed with total contempt, because they had the same reputation as Gypsies – they were considered thieving vagabonds.

In spite of this, it is a fact of life that there are people all over the world, like the Irish, who have an abundance of ethnic and religious prejudices themselves, which they direct at others, and yet these very same people get very annoyed if the prejudice is directed back at them.

But I think even if I had focused on examples of Irish Catholic prejudice and the prejudice that exists worldwide, I would still be annoyed at the Scots because I had convinced myself at the time that the anti-Irish prejudice in Scotland was entirely without justification.

As far as I was concerned the Irish had never done anything to the Scots to merit this animosity. As for myself, I had viewed the Scots people as cousins – right up to the time I stepped off the boat in Glasgow and became aware for the first time of the reality of the relations between the two peoples.

In the months that followed, I remained very curious about the reasons why some, not all, of the Scots had such a negative view of me and my fellow countrymen. One night down in the canteen while I was having a few beers I made an effort to find out about the roots of this prejudice by cross-examining two Scotsmen who were in the canteen with me having a beer.

I might not have questioned them had I been sober, or had I given the matter some thought before I started my cross-examination.

I knew that I had to be very careful however, because it was considered bad manners to go around asking people about their prejudices. It was like trying to ask serious questions about their sex life, because sex and prejudice were taboo subjects.

I thought it was *alright* to question these two particular Scotsmen because I was on very good terms with them – one was an electrician who I helped out when he had an accident in the tunnel; the other was married to a Donegal girl and he talked to me frequently about Donegal.

During my time in the Highlands, I had talked frankly with these men on a wide variety of subjects, but I must admit that anti-Irish prejudice in Scotland had not been one of them.

When I approached them about the subject that interested me so much both acted with a great deal of embarrassment, and both denied that they were bigoted, or that they knew any Scots person who was prejudiced in any way.

Not only did they act as if they were insulted by the very question, but they also believed I was making an outrageous accusation that had no basis in reality. Judging by their reaction, it seemed to be the most outrageous accusation they had ever heard.

I was embarrassed that I had even asked the question in the first place. I thought I had been very naïve to expect an honest answer to this very sensitive question, even from men I was friendly with.

Both came back to me days later to confirm their outrage about being asked the question and it was obvious I had really touched a raw nerve.

They also informed me that it was 'unnatural' for a working man to discuss subjects like that, and that no working man should talk like that unless he was looking for trouble.

I did not see any point in arguing with them or adding fuel to the fire by calling them a couple of liars.

Besides, I did not know how my questions would get twisted as they made their way to the management of the Carmichael Company. For all I knew, the pair of them had put the story all over the camp.

So I backed off and let the issue cool down. I acted very friendly with the pair of them in the weeks that followed, as if nothing had ever happened between us.

A few weeks later I went down to the canteen one evening for a beer and there was the electrician sitting alone in a corner and he appeared to be very drunk. When he saw me, he called me over and asked me to join him. After I sat down, he asked why I had asked the questions about prejudice and why I had directed the questions at him.

I told him that rarely a day passed that I did not hear some unflattering remark about the Irish and that I had been especially infuriated at the man who had told the camp management he would not sleep in any bed in which an Irishman had slept. I said I had asked him the question because I thought we were friends and he would give me an honest answer.

So, I repeated the question: I wanted to know why there was so much animosity towards the Irish and what the reason was for it. I said I was just curious.

He didn't answer for a time, but then he told me the reasons that he knew about – reasons that made sense to him. I listened quietly, not wanting to distract him with questions. I thought this might be the only chance I would ever have to get my questions answered. He said the Scots disliked the Irish because they took all they could from the British and gave nothing in return. He said they were seen as a treacherous people who could never be trusted. He said they were viewed as uncivilised and ignorant and their religion was like witchcraft. And he said his parents told him that all sorts of despicable behaviour went on in convents between priests and nuns.

Finally, he said that the Pope was not really a religious man, but was a power hungry tyrant who was trying to overthrow governments all around the world, and that the Irish knew this but followed him anyway. He said most of his friends did not really think Catholicism was a religion.

While he was spilling the beans, he kept looking at me with a mixture of apprehension and bravado, as if he was expecting me to jump him any moment, or get some of my friends to jump him.

When he was finished he stared drunkenly at me to see what I would say, but I did not confront him. Instead, I just thanked him very much for the information and got up and walked out of the canteen.

The information he had given was not a total surprise. I had read in the Irish papers a few years earlier how an anti-Catholic book, supposedly the memoirs of a nun who had left a convent and converted to the Presbyterian faith, was circulating in Protestant communities in Northern Ireland.

Supposedly, the book was full of graphic sexual details about what went on behind the convent walls. While many Protestants seem to accept the book as a portrait of the truth, in Donegal the Catholic community viewed it as a piece of Orange propaganda.

The attitude towards the Pope was also familiar to me because I knew that one of the war cries of militant Orangemen was – 'To Hell With the Pope,' and that, too, could have travelled across to Scotland.

It occurred to me as I left the canteen that this nun's book had probably found its way over to Scotland. If it enjoyed a wide circulation it could be partially responsible for the Scots' animosity towards the Catholic Church.

I believed at the time that all the sexual allegations made against the Irish Catholic clergy were without foundation, but that was long before the sexual scandals forty years later that were to lend some credence to the Scotsmen's opinions.

As for the 'ungrateful Irish' accusation this was a chicken and egg situation: the Irish were angry at the prejudice and felt they believed they had no need to be grateful to those who despised them. This in turn generated fresh animosity among the Scots. It was a vicious circle.

The information given to me by Gordon did not make me feel any better about the bigotry to which I was being exposed, although it did give me an understanding of why they behaved in a manner that seemed irrational to me. Their bigotry was the result of the way they were brought up and they would have difficulty in behaving in any other way.

This information also explained why the Scots wanted no part of any idea that they had a Celtic Irish heritage. With an attitude like that towards the Irish, they would consider it an insult to imply that they shared common roots with the Irish. And this explained why they emphasised their French (Norman) connection, instead of the Gaelic Irish connection.

I did not see Gordon for a few days after our little conference and when I did run into him he was very embarrassed. I pretended not to notice this and was as friendly to him as if nothing happened.

A couple of times he seemed on the edge of saying something to me, but he would always change his mind and his thoughts were not expressed.

*

I worked with Gordon for a year after that but the subject of bigotry never came up again. I did notice two changes in his behaviour, however.

Before our little chat, he used to make mildly insulting remarks, always in a joking tone. After our chat he never once made a joke about the Irish.

The second change was a total surprise to me.

Shortly after our talk, I went home to Ireland for a week's holiday along with most other men in the camp. One day I walked into Brennan's Bar, which was across the street from Campbell's Hotel, and there was Gordon bellied up to the bar having a fine old time of it drinking with other tunnel tigers. Later that night, I heard him joining in the singsong, giving a credible performance of some traditional Scots songs.

When we got back to Scotland, I did not ask what prompted his visit to Ireland, but he must have enjoyed himself because he came to Donegal every year for years after that.

As for me, I realised I could do nothing about this deep-rooted prejudice, and so I ignored it as best I could.

But there was one direct result of my conversation with Gordon: I knew it would be very difficult to settle down in Scotland and be exposed to ethnic animosity for the rest of my life, no matter who was to blame for it.

It was out of the question.

CHAPTER 15

Religion?

ON THE SURFACE RELIGION seemed to be an issue between the Catholic Irish and the Protestant Irish who worked in the Highlands, but if you observed both the Catholics and the Protestants closely, you would never see the differences between the two communities expressed by any public display of religious worship.

Indeed most members of these sects rarely left the camp to go to church on Sunday mornings.

The lack of enthusiasm among Donegal Catholics for Sunday Mass while in Scotland was remarkable when one considers that back in Donegal everyone who could walk at all went to Church on Sunday. People who did not attend church were considered very eccentric.

The odd thing about the behaviour of these Catholics was that they had helped build the new Catholic Church in Pitlochry, but once it was built not everybody felt the need to go to services there every Sunday. Indeed, the lack of attendance at Mass was considered unimportant and nobody even talked about it.

The same situation existed in the 'Protestant camp' a few miles away. Men who were ferociously dedicated to various Christian sects never thought it important to go out on Sunday and find a church of their own denomination. Only a small minority were devout churchgoers.

Like their Catholic co-workers, the Protestants were content to relax around the camp, or go off to Perth or Glasgow to visit the bars. If they felt any guilt about this they certainly did not talk about it.

Religious affiliation was obviously more a cultural or tribal identity than a firm belief in the merits of the doctrine of any one particular Christian sect. And since neither tribe went to church very often, it was very hard for German or Polish workers to understand what the Protestants and the Catholics were quarrelling about. Both sides seemed irrational to them.

There were a number of evangelists who came into the Protestant camp from time to time in an attempt to 'save souls', but they rarely drew a crowd. In fact, sometimes they were met with hostility when they invaded huts and tried to persuade men to listen to them.

One of these evangelists showed up at Dalcroy one Sunday afternoon and started to preach at the main gate. He quickly gathered a group of men who wanted to find out what he was preaching about.

At first the men who stopped to listen to him thought that the man was confused – that he thought he was in the Protestant camp. So they pointed out his error to him.

But the preacher assured them that he had not made a mistake and that he wanted to harvest Catholic souls. When his small audience heard this most of them departed.

The preacher was there about half an hour when I came on the scene. I could see the lack of interest in his message, although he did not appear to be discouraged.

As I listened to him I remembered an incident I had witnessed in the Square in Donegal Town, when a preacher from Derry had given a public sermon for more than an hour, while he was being ignored by the hundreds of shoppers who walked by. He might have been invisible as far as they were concerned

He kept saying that he knew that nobody in the Square wanted to hear about Jesus, but he was going to tell them about Jesus anyway. I knew there had to be a large number of Protestants walking around in the Square that day, but they, too, had ignored him.

One thing that I gave Protestant evangelists credit for was that many of them were outspoken about their religious beliefs. They had a passion for their faith that seemed absent in Catholics, who went to a Latin Mass that they did not understand, and who found it embarrassing to talk about their religion in public.

Some of the working men who were of German or Polish origin

were bewildered by the reasons for the divisions between the two Irish religious groups.

As far as they could determine both groups were very much alike – so much alike that they thought that it was impossible to determine who was a Protestant and who was a Catholic simply by observing them.

This was an opinion that would have been instantly disputed by both Catholics and Protestants, who believed that there were very clear physical differences between the members of each group.

The expression 'very Protestant looking' was often heard in the camp at Dalcroy.

The Protestants used the expression 'looked like a papist' to describe someone that they thought had typical Catholic physical characteristics.

The reality is that it was impossible to determine religion by physical appearance, since neither group had distinctive skin or hair colouring. Nor was their stature in any way different. But you could not convince members of either group that this was so.

There were differences in the behaviour of the two groups, and based on that behaviour it was sometimes possible to tell the difference between Catholics and Protestants.

For instance, Donegal Catholics rarely socialised with Scots working men, while Ulster Protestants socialised with them frequently. So, if you saw men with Irish accents out at a bar with a group of Scotsmen, it would be safe to assume the Irishmen were Ulster Protestants.

The mannerisms of both groups had subtle differences, which were often an indication of religious affiliation.

The Protestants were often outspoken and felt no need to hide what they were thinking. They were inclined to ask questions bluntly, and they were not given to smiling very much.

The Protestants also had a direct way of looking at people. It was as if they were psychoanalysing them. Making direct eye contact was one of their trademarks.

Donegal Catholics were 'friendly' and easygoing, and unlike the

Protestants they thought it very bad manners to question people directly or be too outspoken when giving opinions. This set them apart from the Protestants.

Another difference was in the names. The Donegalmen had names like Sweeney, Boyle, Gallagher, O'Donnell, Bonar – all the Gaelic Irish names that have been in Ireland for a thousand years. The Protestants had the surnames they had brought across from Britain hundreds of years ago, names like Stuart, Graham, Gordon, Trimble, and many other Scots and English surnames.

Of course, there are exceptions to every rule. There are many Protestants with names like Sweeney and O'Donnell, and there are Catholics with names like McClean, Hunter or Campbell.

The real difference between the two groups lies in their political affiliation, not in their religious affiliation. Ulster Protestants see themselves as British citizens who give all their loyalty to the British Crown. Catholics in Northern Ireland see themselves as citizens of Ireland and they would dearly love to rid themselves of the British Crown. Donegal Catholics are sympathetic to that view.

But the divisions are more complex than that. Beyond the political differences, there is a perception that each belongs to a different 'tribe' and affiliation with that tribe is for life. A Protestant can become Catholic and turn his back on the Crown, but there will still be those in the Catholic community who judge him on his past affiliations, not his present ones.

A Catholic who sides with the Protestant cause is distrusted by both sides.

Up in the Highlands in the Fifties, the two tribes led separate lives and had very little to do with each other. Fifty years later, the same can be said about the Catholics and Protestants of Ulster today. Some things never change.

CHAPTER 16

Tommy Lavelle

AS A RULE, Ulster Protestants did not associate with Irish Catholics in the Highlands of Scotland, but there is an exception to every rule and in this case the exception was Tommy Lavelle.

I first met Lavelle when I worked as a powder monkey on the site at Dalcroy. Lavelle was driving a digger, which was an excavator that was used for loading trucks with rocks and sand.

The rocks and sand were debris from the tunnel, which was piled near the excavator whenever the rock crusher had an abundance of material coming out to keep it busy. The skips were then diverted to a siding and the excess was dumped in a huge mound for use by the crusher at weekends when there was very little debris coming out of the tunnel.

Since I had a great deal of spare time on my hands, I sometimes went down and watched Lavelle operate the excavator. The power and abilities of this machine fascinated me.

Lavelle was a master at operating the digger, and also adept at operating a bulldozer and a dragline, and he could control these massive machines as if they were toys.

Lavelle was the first to initiate a conversation and I discovered he had an outgoing and friendly personality. I also discovered he had spent holidays near Dungloe, and this gave us plenty to talk about. He seemed genuinely delighted to talk to me about north-west Donegal.

Our friendship developed slowly at first, mainly because I did not feel that comfortable with getting too close to an Orangeman. I had become so gun shy with Ulster Protestants that I had difficulty believing that there might be some of them who were not out to be insulting to Catholics all the time. But I learned that Lavelle was just a friendly person with no animosity to anyone.

After a few weeks he began to tag along with Heinz – a German friend of mine – when Heinz and I went into the bars and dance halls in Dunkeld. Although Heinz was a Lutheran he was very cold at first to Lavelle when he heard he was a Protestant.

'You are one of those British Irishmen aren't you?' said Heinz. 'You line up with the British to fight the Germans.'

'I'm not a *British* Irishman, I am an Irish Irishman,' Tommy said. 'And anyway, if I remember right there was plenty of Paddy Campbell's tribe fighting for the British too. All volunteers.'

I thought there was going to be a quarrel but both men decided the game wasn't worth it, and they never discussed the war again.

By the end of the night the easy-going nature of both men took control and they developed a very good relationship.

I got a little nervous when Tommy started talking about Irish Catholics being involved in the war against Hitler, because I had not told Heinz about my brothers serving in the war . . . and I had told Lavelle.

Fortunately, Tommy said nothing about my brothers and I was spared any embarrassment that the revelation might have caused. But I decided there and then I was going to tell Heinz all about my brothers at the first suitable opportunity.

Both Lavelle and Heinz were relentless in their pursuit of women and it was relatively easy for them because they were such mature, self-confident men who had a great deal of experience. I had just turned nineteen and was shy and I did not try too hard because I had it in my head that Scots girls would not go out with

Irish Catholics. One Saturday night Lavelle introduced me to this pretty girl from Dunkeld and when she seemed so at ease in my company I invited her to go to a dance the following Saturday. She accepted.

On the way back to the camp in Lavelle's car, Heinz seemed very quiet. Then he turned to me and asked me why my date thought I was a Protestant.

I was puzzled by the question. I told him we had not talked about religion.

'Well, she told me she was delighted to meet a nice Protestant boy from Ireland. So, she thinks you are a Protestant.'

I looked at Lavelle and asked him if he'd told the girl I was Protestant. He said he had.

'I wanted to get you a date. She need never know unless you tell her.'

I was surprised at Lavelle's action, but not astounded. Lavelle put little value in religious affiliation himself. He told me that when he lived in Glasgow, where there are a lot of girls of Irish Catholic background, he had often posed as a Catholic just to get the chance of going out with them. And that he had even attended Sunday Mass with some of the girls.

The trouble with Lavelle was that he did not see why everyone did not share his views, and he saw no harm at all in giving me a new religion for the following Saturday night.

'This religion business is a lot of nonsense,' he said. 'It is madness. I had an uncle who married a Catholic and the rest of the family outlawed him – all of them except me. He moved to Dublin to get away from it.'

Lavelle went on to say that his family would not be happy if they knew he was hanging around with me.

'Even my wife would disown me,' he said.

I looked at him to see if he was joking when he mentioned a wife, but he was not smiling.

'I didn't know you were married. You are going out with all these girls'

Lavelle laughed and said I was very innocent – straight from the bogs. He said he had a wife and four children and that he was very good to them. He took good care of them. What he did over in Scotland was his business. Lavelle turned to Heinz and asked him if he were married. Heinz told him to mind his own business.

Nonetheless, I had a real problem with Annie, the Dunkeld girl. I told Lavelle I would not go over there to see her the following week unless she knew about my religion.

I told him that telling the girl I was a Protestant was not fair to either one of us. Lavelle saw I was serious and called up one of her friends the following day and explained the situation. She called the date off, but I found another Scots girl the following Saturday night who did not mind that I was a Catholic.

*

The more time I spent with Lavelle, the more I realised how different we were in many ways.

For instance, I was inflexible when it came to my religion or my ethnic background. He was quite flexible and could adjust to any new situation.

He said that he visited New York once, and he never mentioned his Protestant background, but instead gave the impression he was 'a Catholic'.

'It was much easier,' he said. 'It is a lot of bother to be an Orangeman in the United States. Nobody understands. They know all about the Catholics, but they think the Protestants are a strange people. I am a practical man. It is not just worth the bother.'

I could understand the logic of his comments, but could not understand how he could be so casual about his heritage, especially since he told me that he always marched in the Twelfth of July parade – which celebrates a Protestant military victory over Catholic forces in 1690.

'They are the greatest fun,' he said. 'I would never miss them.'

There were times when I thought Lavelle was only trying to tease me when he made remarks like that. But most of the time I knew he was serious, since he repeated the stories from time to time.

I had difficulty with his cheating on his wife, too. I was amazed that he did not see anything wrong with it or that he was surprised that I appeared judgmental. When I gave out to him about cheating it was the only time that I saw him get very upset. His face got all red and he had an angry gleam in his eye.

'That's what is wrong with you papists,' he said. 'You want everyone to live by the rules laid down by the Pope. We are a free people. We can do what we like. There is no Pope in Protestant hearts.'

It was very hard to get angry at Lavelle because all of his remarks were usually delivered with a smile. The man was incapable of harbouring any real animosity.

Even so, it was obvious to me we marched to the sound of a different drum.

*

About a month after I met him Tommy he told me that I should not be working with a pick and shovel ... that I should be driving an excavator like him.

'Easier said than done,' I said. 'Even Andy Campbell would not give me a job like that, since I don't know anything about it.'

Lavelle then told me that if I were really interested he would get me a job as his helper, since the helper he had, a Scotsman, was leaving at the end of the week. Then he would teach me how to operate the machine.

I told him I was interested, but I did not see how he could manage it.

It was very hard to get into the 'Black Gang' – the name given to plumbers, auto mechanics, and the drivers of heavy equipment, who were often black with grease. But Lavelle must have been very influential because the following day he came to me and told me that the job was mine if I wanted it.

I was delighted because I viewed this apprenticeship as a great opportunity. Digger drivers were always in demand and the salary was excellent. It was considered a skilled trade.

The following Monday I began my career as a helper on the digger. I liked the job from the very beginning.

When I told the other men on the tip and in the tunnel that I was going on the digger as a helper, there was no rush to congratulate me. They were just curious about how I got such a job. When I told them who got it for me, there was silence.

Lavelle's machine was immense and it could scoop up a truckload of dirt in one scoop. It had a whole cabin full of levers and pedals that controlled the boom which supported the huge bucket that scooped up the material. Tommy operated all the controls with the greatest of dexterity, like a skilled gambler dealing cards. I was fascinated by the whole operation.

My job was a simple one. I had to grease all sections of the machinery periodically and I had to keep an eye on the diesel to make sure there was always enough in the tank. When those chores were completed, I was free to watch Tommy doing his magic on the levers.

My lessons on how to drive a digger began almost immediately. First, I was introduced to each lever and was told its function. Next, I was given a demonstration on how it worked.

Next, I was shown how certain levers worked in tandem with others: the lever that controlled the movements of the cabin in a semicircle was often used in combination with the lever that lowered or raised the boom. The lever that lowered or raised the boom was frequently used in combination with the lever that controlled the bucket, which could be moved backward or forward.

When I was able to identify all the levers, Tommy coached me carefully on how to operate each one of them, one at a time. Then came the really hard part: using them in combination.

In a way, operating the digger was very much like driving a car that had a stick shift: using both hands and both feet to manipulate the clutch, the gas pedal, the stick shift, and the steering wheel, all at one time.

In the beginning, the learner will stall the car, send the gears screeching, and forget where the brake pedal is, but after months of practising the whole thing becomes very easy and one can drive without any stress.

The digger proved a little more difficult to learn. But since I had a very patient coach and plenty of time, I became adept with the levers and could pick up buckets of gravel with great ease.

For months I practised by picking up buckets of gravel and

then dumping them in a heap near the digger; I was very careful and slow at first, but gradually I picked up speed.

The final stage of my coaching began when Tommy announced one morning that he was going to allow me to load one of the trucks. I was thrilled.

A few minutes later a truck pulled up and Tommy got out of the cab and told me to load the truck. As soon as the truck driver saw me behind the controls, he jumped out of the truck and told Tommy that he certainly was not going to stay in the truck while someone like me came swinging over it with several tons of rock and gravel.

He said Tommy was a lunatic to allow me behind the levers and that he ought to be locked up.

Tommy told me to pay no attention to him, and I very carefully dug the bucket into a mound of gravel and then lifted it clear of the mound. I swung the cabin around to face the back of the truck and gently lowered the bucket into the back of it. Then I released a bolt that opened a trapdoor at the bottom of the bucket. As I lifted the bucket out of the truck, the trapdoor opened and the material rolled gently into the bed of the truck. No problem.

Tommy was very proud of how well the lesson went, and so was I. But the truck driver, who was a Scot, was furious, and he told Tommy that letting people like me operate a huge machine like that could get somebody killed.

I was allowed to fill three more trucks. Each time, the truck drivers headed for safety, much to my annoyance, but I filled all the trucks without incident. I was very proud of my achievement.

However, towards the end of the day, I saw Scott, the chief of the Black Gang, coming down towards the digger and right away I sensed there was going to be trouble.

He ignored me and went over to Lavelle and asked him why I had been driving the digger. Tommy told him that I had been trained and was capable of operating the machine safely.

'But why have you trained him? You have never trained anyone else.'

'Well, he is a cousin of Andy Campbell, and I thought Andy might like it,' said Lavelle.

Scott stared at him for a moment and then walked away.

Tommy sniggered when Scott left. He said he knew Scott would do nothing if Andy Campbell's name were brought up, because Campbell had more power than Scott had.

'The Carmichaels need the Campbells and they have a great deal of respect for them. Andy's word is the law here.'

I did not know what Andy would think of me operating a digger, but the issue never surfaced.

I thought he probably knew about it anyway since he seemed to know everything else about my activities.

During the next few months, I operated the digger almost every day. I became so competent that some of the truck drivers even got up enough courage to remain with their trucks.

However, the majority of them still headed for the hills whenever I was at the controls.

I really did not blame the truck drivers for feeling nervous. One small error on my part could send several tons of rocks down on the truck cabin. Or I could hit the cabin with the heavy bucket and turn it over. So their caution was understandable.

Eventually, however, my relationship with Lavelle did create problems, but it was not with Scott of the Black Gang, or the truck drivers. It was with my own Irish Catholic co-workers, who objected to me hanging out with Lavelle, whom they called 'that black Protestant from the North.'

The anti-Lavelle sentiment built up in my hut over a six-month period. At first there were some remarks made about the fact that I was going out to the bars with Heinz and Lavelle.

I paid no attention to that, because I did not see that it was anybody's business who I went out with. Nobody had started up a fuss when I first went out drinking with Heinz.

The objection raised initially seemed to be that I was selecting these two as friends and made no effort to fraternise with some of the young people who came from my parish, which was not true.

My response to that was that Heinz and Lavelle were fun to be with, we got on well together and that I did not see anything wrong with that. Finally, I told them it was none of their business.

Then the focus was turned on Lavelle alone . . . he was a bigoted Orangeman and I should not be hanging out with someone like that. I told them that Lavelle was not a bigot – that he had no time for bigots and that they just didn't know him.

There was no heat in any of the remarks made; all were delivered in a conversational tone, so I did not take them seriously.

When I got the job on the digger with Lavelle, the temperature of the remarks got higher. Now, it was being insinuated, with a smile, that I had sold out to the British and that maybe I should go off to live in the Protestant camp and maybe even become a member of the Orange Order.

At this point I realised I had a problem. Behind the smiles there was some very real animosity.

From my point of view I was not doing anything wrong. I was socialising with Lavelle because he was good company, and when the digger job was offered to me I would have been a fool not to take it. I also knew that if I had wound up with two good jobs in a row I could thank the influence of Andy Campbell, not Tommy Lavelle.

I did not say anything to Lavelle about what was going on back at the hut, but I found out later that he knew about it all along.

One day when I came into work there was a young man named Hugh Walsh operating the digger. Walsh had been on the night shift but I had never met him until that day.

I asked him where Tommy was, and he said Tommy was on the night shift. They had switched. I asked him why I was not going on the night shift too, and he said the night shift operator did not have a helper.

Walsh was very nice to me and later on in the day he said that he would continue to teach me how to operate the digger.

'There is not nearly enough Irish in the Black Gang,' he said.

When it was close to quitting time I told Walsh that I was going to stay late to see Tommy.

He stared at me for a while and then said I should try to hang out with my own kind and not with the Orange people.

I told him he should try minding his own business, and I walked away from him.

*

Tommy was cool to me and I got none of the usual cheery greetings. When I asked him what happened he told he was ordered to go on the night shift, without explanation. When I asked him if he had any idea of the reason, he said my 'papist buddies' had probably caused it.

He told me he knew all about the talk that had gone on among the Catholic workers.

'What I find ridiculous is that you people are always whining about Orange bigots, but you people are just as bigoted yourselves. If one of us says hello to one of you, you wonder what he is up to.'

That night marked the end of the Lavelle/Campbell friendship. We never socialised again.

*

I was angry and embarrassed by the Lavelle incident. I never found out who had been responsible for him being dumped on the night shift.

When Lavelle had said it was some of my Catholic friends who had come after him, I had assumed that it was Andy Campbell. I was so angry at first that I decided I would confront Andy about it. It was a good job that I did not do this because years later I found out that he had nothing at all to do with it.

Later I came to the conclusion it was probably Scott, the chief of the Black Gang, who decided that – for the sake of peace on the job – we two should be separated.

*

Although Lavelle and I were never friendly after that, our relationship had made a deep impression on me and had an effect for many years afterwards.

During the Seventies and Eighties when Catholics and Protestants were in constant confrontation with one another in the North, I was frequently alienated by the open prejudice expressed by some of the Protestant leaders. I would react badly to it and be inclined to think that all Protestants were like their leaders.

Then I would remember Lavelle and think about his complete lack of prejudice. Using him as a frame of reference I knew that not all Protestants were bigots – that there probably were plenty more like him out there. This effect continued on into the present time.

*

I have never made any effort to find the man I have called Tommy Lavelle in this book. I have avoided pursuing him for a number of reasons.

First, we did not part as friends and I have no idea how he would react to a visit from me.

Then, I was afraid that he might have turned into one of these Protestant extremists during the Troubles and that when I found out I would want to have nothing to do with him . . . or he with me.

Taking everything into consideration, I decided that I would rather remember the easy-going Antrim man as he was in the early Fifties, and not what he might have become later.

However, if a real and lasting peace comes to the North I may try to find him. I know his real name; his age; and the town he came from.

It shouldn't be so hard to find him . . . if he is still alive.

CHAPTER 17

The Germans

ONE OF THE FIRST friends I made at Dalcroy was the young German ex-POW named Franz Heinz. He had a bed near me in my hut and we hit it off right away. I remained a friend of Heinz all the time I was at Dalcroy.

There were a number of German ex-POWs among men of many nationalities working at Dalcroy. The German ex-POWs in the camp stood out for a number of reasons.

First of all, one would imagine that these Germans, who had served in a defeated army and whose nation was being ostracised in Britain in the early Fifties because of the behaviour of German troops during the war, would appear dejected or defensive.

But this was not the case at all; these men looked everyone in the eye and conducted themselves with a great deal of self-confidence. When they discussed Germany's role in the war they never had any apologies to make.

Of the two Germans in my hut, I got to know one of them, Heinz, very well, because his bed was near mine and he was close to my own age.

The second one, Carl, who was much older than me, was one of the most forceful personalities in the building. Carl was already in residence the first night I arrived and I became aware of him even though he was at the other end of the hut.

You could not help but notice Carl's sleeping arrangements. He had twice the space of anyone else and had his area decorated like a home away from home. He always made his bed before he left for work in the morning, and the bed was made with clean sheets and a clean eiderdown. A Catholic, he had a large crucifix hung over the head of his bed.

Photographs were hung on the wall near his bed, a small broom was leaning against the bed, and clean towels were stacked neatly above his locker. Somehow Carl had managed to get rid of the bed next to him and it its place he had put a carpet on the floor, which was immaculately clean.

Carl also had a comfortable armchair in this space, and a small table in the corner, on which he had set up an electric grill.

Every night after he came back from work, Carl would take off his work clothes and place them in his locker. Then he would go out to the shower room and have a warm shower. This was in contrast to most of the other inhabitants of the hut, who showered only once or twice a week.

After his shower Carl would put on a set of pyjamas and a dressing gown, and then he would begin to grill steak and onions, which were accompanied by fresh bread and butter. Occasionally he would have a bottle of wine with his meal, but mostly he had a mug of tea. While he was dining, he would listen to programmes broadcast from Germany on the long wave radio.

Carl had no hesitation in telling the other workers in the hut that he thought they lived like pigs. He would say that they did not have to live like pigs.

He would deliver his condemnation in a conversational tone that was not confrontational. Nobody seemed to get angry with him, even though they did not follow his advice.

The first time I heard Carl talk like that I was inclined to agree with him. When I arrived in the hut that first night, I was aware of the rancid smell of dirty feet and unwashed bodies. I wondered how the men could put up with it.

But after a time I got used to the smell and I did what most of the others did: I washed my hands and my head and my chest every night and I only got into the showers when I became aware of the smell of my own feet.

Carl knew that the men in the hut would love to have steak and onions every night, but they were either too cheap to buy the meat or too tired or lazy to prepare it after a long day's work.

So, he never offered to share as he ate his meal every night, while the other workers did their best to ignore him.

In spite of the fact that it was the Scots that had the reputation of being cheap, the Irish workers showed an even greater reluctance to buy beef, and most of them would only consume it if they got it free, or if poachers shared the product of the hunt with them.

Instead, most of the Irish workers went down to the mess hall and ate the gut-churning dinner provided by the company. Then they went back to the hut and tried to digest it as best they could.

Carl liked to tell his audience that some of the middle-aged ladies who worked in the mess hall practised the oldest profession in the world in their leisure hours, and that nobody should eat food cooked by them. But his advice fell on deaf ears.

I observed Carl's routine for a couple of weeks, and after I got my first wage packet the smell of the steak proved too much for me. So I went to the truck that came up twice a week from Pitlochry and bought some steak and onions. I then borrowed a pan from one of the men and waited impatiently, with my mouth watering, for the steak to cook.

The first bite I took tasted so bad that I thought the meat was rotten, since it certainly did not taste like any steak I had ever eaten in Ireland. But one of the workers told me the steak had been imported from the Argentine and that it did not taste at all like Irish or British beef. However, the men said I would acquire a taste for it in time, but I knew this was not going to happen.

This was all the beef that was available in Scotland at that time, because the British were importing meat to build up their own herds, which had become depleted in the war. I decided that the slop in the dining room had more flavour than this import, and I did not envy Carl anymore.

*

One of the most unusual aspects of Carl's living area was the number of framed photographs of him in German Army uniform.

Carl must have brought a camera with him to war, because there were poses of him in Warsaw, and in various other places in Russia and the Ukraine.

Many of these photographs showed Carl with young women and the women were always smiling. Carl told anyone who asked him about the smiling girls in the photographs that the girls were smiling because they were happy that he had liberated them from the dictators who had ruled Eastern Europe until the Germans came along.

He used to get some of the Polish workers very angry when he claimed that the Polish Army had not put up a fight and that the Germans just rolled into Warsaw without opposition.

He said the Poles had confronted the German tanks with Polish troops on horseback and run away after the first shots were fired.

There was one Highland Scot and one Polish worker in the hut and they detested Carl. Behind his back they called him a liar and a 'poof' – the term used for a homosexual. The 'poof' tag was earned because of his habit of dressing up every night and his preoccupation with cleanliness.

No one would dare say anything like that to his face, because Carl was a very big man who was obviously fearless.

The Irish workers usually ignored Carl, although the trouble-makers among them often encouraged him to talk about his war experiences just to annoy the Scot.

But in spite of his willingness to talk to anyone who approached him, one never really got close to Carl, because he never talked about his pre-war life and he never once mentioned if he had a family.

Carl was very selective about the subjects he would discuss, and if he did not like your question, you would not get an answer.

Carl's departure from the hut came during a rare spell of hot weather in the summer, when the sun shone every day and the heat outside made the Highlands seem like Miami Beach.

For a few days the temperature soared into the 90s and, as a result, the huts became ovens that trapped the heat inside.

Most of the men handled the situation by leaving the doors

and windows open and sleeping without blankets or sheets, but one man who had a bed across from Carl went further – he took everything off and lay on top off his bed naked.

For some reason, this drove the German berserk. He demanded that the man cover himself like a Christian and not lie around like a pagan. But the man, who was from Donegal, told Carl that he was not going to listen to him and that he would do as he pleased. After two nights of yelling, the German was the first to acknowledge defeat by moving to another hut, where he was restricted to one space.

After Carl left, some of the Irishmen went to the nudist and told him to find a loin cloth or move out. The nudist covered up rather than risk a confrontation with the other Irish workers.

*

Heinz, the second German in the hut, seemed pleased that Carl had gone, because he was embarrassed by him. Unlike Carl, Heinz had an easy-going personality that enabled him to get along with most people.

I once asked him why he seemed to dislike Carl, and he said that the war photographs on display were pointless and were just designed to provoke the British. He said that although he would love to get even with the British in another war, Carl's tactics were childish.

Sometimes I went out to the bars with Heinz, and when he had a few drinks he would get very bitter about the German defeat in the war. He would say that it took the whole world to beat the Germans, and that if the Americans had stayed out of the war Germany would have ruled the world. He said he could never understand why the Americans entered the war in the first place.

As far as Heinz was concerned Germany had been the innocent victim in the conflict. He blamed the Jews for getting the Americans involved, and he claimed the Jews had also betrayed Germany in World War 1.

However, by this time, in the early Fifties, the full story about the horrors of the slaughter in the concentration camps was beginning to emerge. While I did not want to begin a big argument with Heinz over this issue, I felt compelled to bring it up, and so I asked him about the stories about mass murder in Belsen and other camps.

At the very mention of Belsen he became very upset. He said there had been no mass executions in the camps, that it was all a pack of lies invented by the Jewish newspapers in America. He said he served all over the eastern front during the war and the only camps he had ever seen were prisoner of war camps.

He also said that he was born near Belsen and could assure me that the atrocities attributed to the camp were pure fiction. He said all these stories had been fabricated by the Jews.

I had been told by one of the other Germans that he thought Heinz had been in the SS, and I wondered if he had something to hide – some secrets he would prefer not to reveal. But I did not press the issue.

However, I was not telling all I knew about the camps either, and I had secrets of my own that I did not reveal to Heinz.

These secrets involved my two brothers, who had taken part in the war against the Nazis. Both had volunteered to serve in the conflict, a fact Heinz would have found very difficult to accept, because he always believed the Irish and Germans had a common enemy: the British.

My brother, Bernie, had served as a radio operator in the American Merchant Marine on the North Atlantic run and had twice been torpedoed by German subs. But he survived and later settled in the United States.

My oldest brother, James, had joined the American Army in the waning days of World War II, and was with the victorious army that overran Germany. He remained in Germany for several years after the war.

It was James who told me about the atrocities committed by the Germans. The first time he came home on leave after the war, he had horrific tales to tell about mass murder and unbelievable barbarity in the concentration camps.

I believed these stories because this was something James was not likely to make up, so I was unable to accept Heinz's denial that any atrocities had taken place.

The stories told by James gave me some idea of the type of crimes committed by the Nazis, but it would be decades before I was aware of the full scope of the Holocaust and the role of the SS in it. In the years that followed, while living in the United States, I read a great many books on the Holocaust written by people of many ethnic backgrounds, and it was only then that the immensity of the crimes committed was revealed to me.

Had I known back in the early Fifties of the scope of the Holocaust, I doubt that I would have been so friendly with Heinz.

*

The lack of animosity between me and other Irishmen in the camp and Heinz and other Germans might seem unusual given the number of Irish who had served in the British Army during the war.

But the Irish had experienced such open anti-Irish Catholic prejudice in Scotland that they could only react to that bigotry, and were totally unaware of the true extent of German barbarity.

As a result, the Irish openly fraternised with the Germans, and few of them cared anything about what the Scots thought of them for doing so. As far as the Irish were concerned, the war was over and the German ex-prisoners were doing nothing to merit anyone's animosity.

CHAPTER 18

The Campbells

ALL THEIR WORKING LIFE, the five Campbell brothers put themselves in harm's way by working on dangerous construction projects. They had become living legends on construction sites all across the north of Scotland since the Thirties.

But then luck ran out for one of the brothers; it struck when Paddy Campbell was asleep in his quarters in Garve, Ross-shire and a rapidly moving fire destroyed the building in which he was sleeping, killing him and one of his nephews.

The accident stunned his brothers, since they had been through so much together, and it was ironic that death came while Paddy was in bed and not deep in a dangerous tunnel.

But life went on for the surviving brothers.

*

Andy Campbell, the leader of the family, had been involved in major construction jobs since he was a young man, and gradually he acquired the expertise that made him in demand as a consultant, even on jobs in which he was not directly involved.

Andy was once hired as a consultant on a major tunnel project in Nigeria. The sandy soil that the tunnel was being driven through kept caving in and the workmen were refusing to go inside the tunnel because they were afraid of a total collapse.

Andy's solution was to spray the inside of the tunnel with a fast-drying concrete that sealed the sides and prevented any further erosion. For this solution he was honoured by the Nigerian Government.

Andy also participated in a major British Navy project at Scotland's Scapa Flow Naval Base during World War II. There he helped design and build concrete barriers between islands in the island chain that surrounded the base.

The function of these barriers was to prevent German submarines from sneaking in between the islands and torpedoing the British ships that were moored there.

These sneak attacks had happened several times and the British were determined to put an end to them.

The job was a very complex one and it had to be finished on a very tight schedule.

During the construction of the barriers many British leaders, including Churchill, visited the base and inspected the work in progress.

Andy remembered speaking to Churchill on several occasions. He said that during these conversations Churchill appeared relaxed and easy-going and not the feisty personality that was portrayed in the newspapers.

When Churchill left the base after his final visit, Campbell was told that he and several others were going to receive 'recognition' from the King for their efforts to seal off Scapa Flow.

But this information created a problem for Campbell.

The problem had its roots in the relationship between Ireland and England, which was not particularly good during this period. Just twenty years had passed since the Irish War of Independence and this was a time when the British were abusing the Irish for failing to declare war on the Germans.

The Irish were obsessed by the fact that the British were occupying north eastern Ireland against the will of the majority of the Irish people. This obsession dominated all their perceptions of the Anglo-Irish relationship.

Even so, there was no willingness among the Irish to side with the Germans in the war, because they were well aware of the threat posed by Hitler to all of Europe, but the idea of officially throwing in their lot with the British was out of the question . . . so long as the British occupied Ulster.

It was Andy's opinion that if he accepted some kind of award from the British, he would be viewed back home as a collaborator. This could make it very difficult for his wife and children, who were living in the brand new home he had built for them near Dungloe.

So, he put out the word to British officials that while he was deeply moved and honoured to be considered for recognition, he thought that, given the fact that his family lived in Ireland, he would rather not be considered for recognition at that time.

Campbell never heard another word about an honours list, and he had no regrets about the position he took.

Campbell's action underscored the very complex relationship between the Irish and the English in the 1940s.

On one hand, there was hostility between the two nations; on the other hand, the British Army had countless thousands of Irishmen among its ranks, and all over Britain there were hundreds of thousands of Irishmen, like Campbell, helping Britain with her war effort.

During this era, it would be difficult to find a person in Ireland who did not have a relative who was involved in the war effort on the British side either as a worker or as a member of the armed forces.

Thus, in spite of a great deal of public posturing on both sides, it was very evident to the leaders of both nations that the Irish were very much involved on the British side during the war, not only with contributions of manpower, but also with continuous shipments of food and material from Ireland to Britain.

The Campbells spoke very little about their adventure in Scapa Flow, but instead liked to talk about the great projects they were involved in after the war, and the list of these projects was long and impressive, and involved many of the major hydroelectric projects in the Highlands.

The brothers loved the challenge of something new and enjoyed the sight of a new hydroelectric project developing in the wilderness.

It was only when the job was near completion that they grew restless and began to look around for a fresh challenge.

Old age finally caught up with Andy Campbell. When he saw that he had become too slow on his feet to get out of harm's way when danger threatened on a construction site, he retired to his comfortable home in Meenmore, outside Dungloe, where he played host to old friends . . . and raised chickens.

'I was really amazed when he began to raise chickens,' said Andy's son, Michael. 'When I worked with him he wouldn't let any of the men raise chickens near their huts because he said real men didn't raise chickens.

'I think he was concerned with the image it might create for the tunnel tigers, and no matter how much we argued that we just wanted fresh eggs, he wouldn't let us.

'Then he comes home and starts to raise chickens himself!'

Image was very important to Andy Campbell. When he was in his sixties and still running a major project, he was approached by a man at the camp who was looking for a job.

The man hadn't seen Andy for years, and very foolishly he told Andy that he had got very old-looking. Andy's response was to punch the man on the chin, knocking him out, and Andy then walked away in a fury.

The man did not get the job.

*

One by one the other Campbell brothers also retired as old age caught up with them, and like their brother Andy, they have remained very much part of the folklore of north west Donegal.

The Campbells may not have received recognition from the British Government for their achievements, but the dams and reservoirs they worked on are scattered all over the Highlands, and these are monuments to the contributions they and the rest of the Irish workers made to the development of the great hydro-electric projects in Scotland.

CHAPTER 19

Some Characters

AS I LOOK BACK on my years in the Highlands, there are many men who I remember clearly: I recall what they looked like, and what it was about them that were so notable.

The long distance kiddies were among these, and so were a number of the German workers. There were also other men who stood out in the crowd for one reason or another.

One of these was a Polish foreman known as Jack. He was a tall, arrogant, self-styled Romeo who was in charge of the crew on the crusher. He was forever making derogatory remarks about the Irish.

The Irish workers did not get angry with Jack because they were contemptuous of him. They jeered at him because of his broken English and his tendency to kiss up to the foremen in the Black Gang. Since they had no respect for him, they did not take his insults too seriously.

Jack liked to boast about the conquests he made every weekend. Every Monday morning after he gave a detailed account of what he had done with every women he dated, he would ask the Irish workers present what *they* had done at the weekend. When he was greeted with silence, he would laugh sarcastically and say that the only thing the Irish were good at was drinking.

He also had a lot to say about his war record. He boasted that he had fought the Germans courageously until Poland was defeated, and had then joined an 'elite' German regiment to fight the Russians, whom he hated more than he hated the Germans. Jack thought there was nothing wrong with joining the German Army after the Polish Army had been defeated – indeed he thought it a very sensible thing to do, given the situation.

According to Jack he had covered himself in glory fighting the

Russians on the Eastern Front and had been awarded many medals by the Germans for his exploits.

Jack was full of praise for the fighting capabilities of the German troops. He had no respect at all for the Russian troops, and he ridiculed the Italians, whom he said could not fight at all.

Jack did not produce any of the medals that he said had been awarded to him, and his explanation for this was that he had to throw the medals away after the war in case the Russians or his fellow Poles discovered them. He said that he had barely escaped with his life as he fled into Western Europe after the war.

The Irish workers did not care whether Jack was a great lover, or whether or not he was a highly decorated war veteran. All they knew was that he was a boring pest and they were sick listening to him.

Most of the time Jack knew better than to pick a fight with the Scots, but from time to time he just could not help himself and he would get involved in a donnybrook with one of them.

On one of those occasions Jack got into a fight with a Scots mechanic over his unflattering remarks about the war record of the British Army.

He said the Germans were twice the soldiers the British were, and if the Scots abandoned the skirts they wore and fought like men then maybe he would consider them real soldiers.

Jack and the Scots mechanic got into a brawl, but the supervisors on the site didn't think the incident funny and Jack and the Scotsman were told to go the office and pick up their final wage packets.

Both pleaded to be allowed to stay on, and both were allowed to stay – on the understanding that there would be no more boxing matches, a condition Jack meekly agreed to.

But Jack's undoing came one Saturday night when Stuart, a Scots electrician, spotted him in a bar in the suburbs of Perth with a blonde woman. The woman was older and did not look like the young women Jack had been seen with up in the Pitlochry area during the week.

Stuart sat in a corner of the bar keeping an eye on Jack until the foreman left the bar with the woman. The electrician then followed the pair until they went into a council house on the outskirts of Perth.

The thing that intrigued Stuart most was that Jack and the woman were behaving like an old married couple who did not have a great deal to say to one another. They just sat there drinking beer all evening, rarely talking. Certainly there were no romantic gestures involved.

Stuart made some inquiries around the bar and discovered that the woman was indeed Jack's wife, and that the pair had come from Poland as a married couple on the eve of World War II and had been settled in the Perth area since then.

His informant also told him Jack's wife often came into the bar during the week with other men when Jack was at Dalcroy. The woman obviously had a greater romantic interest in these men than she had in Jack.

Stuart was all set to reveal this information up in Dalcroy the following morning. He was looking forward to embarrassing Jack, but late Sunday night after Jack arrived back in the camp one of Jack's girlfriends, a young woman from nearby Aberfeldy, had come into the camp looking for him. When she found him she read the riot act to him in front of a score of people because she discovered he was seeing another young woman in Pitlochry.

Among the insults she threw at him was the accusation that he was not a 'real man' – that he wanted women to treat him like a little boy because he saw all women as mothers.

This scene must have embarrassed the arrogant Jack no end, since the confrontation was held in public. The following morning he quit his job and fled the camp, unable to live with the smirks that were evident on the faces of all those who witnessed the confrontation.

It was just as well for Jack that he left because Stuart would have dropped a second bombshell, and Jack would not have been able to face his workers if his non-existent war record were also made public knowledge.

Jack would have received little sympathy around the camp had he stayed and the only thing that the Irish were sorry about was that he had ran away, because they would have loved to snicker at him, and talk sarcastically about fake war heroes and mama's boys.

But nobody forgot Jack, or how he met his Waterloo in the Highlands of Scotland.

*

A young Irish labourer named Danny was another tunnel tiger that few forgot.

Danny was a clean cut, clean living young man who stood out among the other young men because he did not drink or smoke and appeared to be deeply religious.

Danny not only went to Mass every Sunday, but he was always handing out pamphlets after Sunday Mass about the Children of Mary or the Pioneer Total Abstinence Association, a group that lectured the Irish on the evils of liquor.

Danny's efforts to get Catholics to attend Mass or to give up the evils of booze met with very little success. The workers liked Danny, and deep down they respected the message he was delivering, but most of them did not consider it a sin to miss Mass, and those who drank were not prepared to give it up, so Danny's sermons fell on deaf ears.

Danny was always neat and tidy and it was amazing how he could keep himself presentable even while working with a pick and shovel. He was also very well organised when it came to money. He rarely spent anything, and it was rumoured he had a bank book with a healthy balance.

Danny was engaged to a young girl back in Ireland. He wrote to her several times a week, and was always receiving letters back from her.

He planned to move back to Ireland and open up a general store out in the countryside in an area where one was needed. His future wife wanted no part of living in Scotland and that suited him, since he was looking forward to settling down in Ireland.

Indeed, Danny seemed to have everything going for him, so it was a major sensation one Monday morning when word came to Dalcroy that Danny was dead – that he had fallen from or was pushed off a bridge in Inverness and had drowned in the river.

Most men who knew him could hardly believe that such a thing had happened. It was one of those deaths that posed a thousand questions, and there was never any answer.

The complete details of Danny's death were never fully determined. He had been seen sitting on a safety wall on the bridge shortly before his death, and whether he then accidentally fell off or was pushed off remains a mystery.

Those who knew Danny knew he had not been drunk, and the police said there were no witnesses to the incident, so one could only speculate about what had happened.

Danny had a considerable amount of money on him when he died and a Scots bank book with seven hundred pounds in it, so whatever had happened was not related to money.

There was a great deal of grief in the camp over the death of Danny. Many of the men even went to Mass the following week as an expression of sympathy.

His family came over to take the body home, and it was rumoured that his fiancée had taken the death badly.

For years afterwards the men in the camp talked about Danny, trying to make some sense of the untimely death of a young man who had touched so many lives. Yet the years passed and they never did make sense of the incident.

*

Another personality who is fresh in my memory was a young man named Sean who had gone to school with me in Dungloe and then wound up with me in the Highlands.

Sean was an easy-going young man who liked to chase girls around Perth and Glasgow on the weekend.

Although we had been close friends back in Donegal, we had

drifted apart a little in the camp because Sean worked on a different part of the site and lived in a different hut. He also went away to Perth every weekend, and I had no interest in doing that.

I had been up in the Highlands for more than a year when a man from Dungloe named Pat Boner came over to me and asked if I was a friend of Sean's. I said I was.

'Well, you should talk some sense into him because he is going to marry a Scots prostitute next weekend,' he said.

I thought at first that this was a bad joke, but when I realised he was serious, I went off to find Sean.

I found him in the canteen drinking beer with a young man named McHugh, who was trying to persuade him not to go ahead with the wedding. McHugh was giving Sean a score of reasons why he should not get married, and Sean listened to all of them. McHugh and I then talked to Sean for hours but without any success.

Sean was very polite and very good-natured in spite of the pressure we were putting on him, but he said he was deeply in love with the girl, and that nothing anyone said would change his mind.

Sean said he was well aware that his fiancée had worked the streets of Glasgow, but he said she had given up the activity years ago and was now very respectable.

He said that he had not lived the life of a saint himself and that he was in no position to judge anyone.

We tried to tell Sean that once a girl got a reputation like that it would follow her all her life.

We pointed out to him that people back in Dungloe were contemptuous of children who were born out of wedlock and did not consider them suitable material to marry anyone.

We also pointed out that there were families in the Dungloe area who were supposedly descended from tinkers, and even though these families had not been on the road for generations they were still viewed as second class citizens.

Given that, he should know well what the people back home would think of a prostitute or the children of a prostitute. He could never go home.

Sean listened politely to all our arguments but in the end he said he was in love and that he did not care what price he had to pay for the marriage He said that anyone who passed up an opportunity to marry someone they loved would regret it all their life. He said he would marry her; he would not bring her home to be insulted; and that he was prepared to live in exile with her for the rest of his life.

That was the way we left it with Sean. A few days later he vanished from the camp and I never saw him again.

Before I emigrated to America I heard he had married and gone off to live in Australia. Others said he had disappeared into Canada.

In later years, although I was curious to learn what had happened to him, I did not ask his family in Donegal about him, because I did not know if they knew about his wife. I thought I had better let sleeping dogs lie.

Sometimes I think about Sean and wonder what became of him. Sometimes I believed he was very unwise to marry the girl he was in love with.

But at other times I wonder if maybe he had done what was best for him. I realised I knew nothing at all about his wife, and had no business at all making judgments about her.

If this Scots girl had left the streets behind her and was forgiven in the eyes of God, who was anyone to judge her?

But then I would think of the terrible social sanctions attached to prostitution and come to the conclusion that it is a brave man who will defy them. Or a very foolish man.

Or a man very much in love.

CHAPTER 20

Outsiders

EVERY COUNTRY IN THE WORLD has a segment or segments of the population that are considered outsiders. They can be considered aliens because of their religion, like the Jews, or aliens because of their colour, like the blacks or Pakistanis. Japan has a Korean minority which is refused citizenship even though numerous generations of Koreans have been born in Japan. Ireland considers blacks, Muslims, Gypsies and tinkers as outsiders, and homosexuals are barely tolerated even though they were born in Ireland.

The Scottish Highlands had more than its share of outsiders during the Forties and Fifties because of the nature of the construction work, which attracted people from all over Europe, and indeed, during the Forties and Fifties the 'outsiders' made up a considerable percentage of the population in some areas.

What was the definition of those who were considered outsiders in Scotland during this period? Poles, Germans, Irish Catholics, Pakistanis, Englishmen, Welshmen, or Gaelic-speaking Scotsmen from the Hebrides. In other words, anyone who was not an English-speaking Scots-born Protestant citizen from the Scottish Lowlands.

Scottish antagonism towards outsiders hardly made the Scots unique among the people of the world; indeed their attitude was typical of the behaviour of other nationalities, including the Irish. However, knowing that there was nothing 'abnormal' about ethnic or religious prejudice did not make it any more acceptable to people like me who were the targets of this prejudice.

One of the Scotsmen that I was friendly with in the Highlands was a young man named Norman MacDonald, who was a Gaelic-speaking Catholic from the Hebrides.

One day I told Norman that I saw no future for me in the Highlands and I was thinking of heading for America.

When I told him this, he revealed that he, too, had plans – that he intended to go to Edinburgh and join the police.

'It is one of the few professions where I will not be picked on because I am from the Islands, or because I am Catholic,' he said.

MacDonald and I used to talk a great deal about the fact that both the Irish and the Islanders were outsiders who were not accepted by other Scots. And he was as annoyed as I was at the open rejection.

He said the Islanders were viewed in the Lowlands as low class people who were barely civilised.

I had no idea how the Lowlanders came to that conclusion because all the Islanders I had ever come across were friendly and completely civilised.

I felt sorry for him since he was being treated as an outsider in his own country.

But MacDonald never reacted openly to this rejection, but maintained an easygoing disposition and was friendly to everyone he met.

In my heart I knew that MacDonald was approaching the problem in a mature manner, because he knew he could not change minds by getting upset. But his mature example was something that the Irish were unable to follow, because they were all too willing to fly off the handle at the slightest provocation, which only made the situation worse.

*

Others in the camp had problems similar to those of the Irish and the Islanders. Among these was a male nurse who tried to be friendly with everyone but was rejected by everyone because he seemed effeminate.

Although the nurse was slightly built and was in his mid forties, a story made the rounds that he had molested a strapping six foot tunnel worker who had gone into the clinic for some aspirins, and the man had had a nervous breakdown because of it.

The story seemed more than a little ridiculous, but that did not stop the majority of the people in the camp from believing in it.

Supposedly, on another occasion he had made another worker unconscious with chloroform and then had his way with him, and the man had not been right in the head since then.

The chances of either story being true were very remote given the delicate physical condition of the nurse. But the homophobia among the men in the camp would not permit logic to get in the way of a good story.

There were Irish youths as young as fifteen working in the camp who had adult working papers and these youths were especially afraid of going anywhere near the clinic because of the stories they had heard.

If one of them was forced by a very bad cold to go and get some medication, he would never go there alone but would always be accompanied by friends who acted as bodyguards.

I do not know what the nurse thought of the way he was being treated, but he soldiered on anyway, making the best of a bad situation.

However, he did get some respect following an accident involving a bus and a car that occurred in the main square of the camp. The mishap resulted in injuries.

The nurse was out of his clinic in a shot and he tended to the victims in a cool professional manner, and in one instance probably saved a life by curbing excessive bleeding in a victim who had a huge gash on his leg. He received a letter of commendation from Carmichael for his efforts.

After this the nurse was acknowledged in public instead of being ignored, and even if he could hardly say he had friends in the camp, the previous hostility was never openly expressed again.

*

Although the nurse earned some relief, the workers had other targets for their hostility, including a man who had once been a neighbour of mine back in Donegal.

I first met him in the camp one evening when I was walking towards the canteen. I had just returned from my trip to Dungloe and I saw a familiar face approaching me and I recognised him as an older man named Brendan who had once lived near me in Dungloe until he moved to Scotland permanently eight years previously.

I had not seen him in the meantime, but he had not changed physically at all and I had no trouble recognising him. But he just stared at me when I gave him a warm greeting and he obviously did not recognise me, which was understandable, since I was only eleven when he left.

The minute we began to have a conversation, however, I became aware that Brendan had changed a great deal – he had in fact recreated himself as a whole new person who was a stranger to me. Brendan had in fact become a Scotsman, complete with thick Scots accent and a slightly condescending tone, as if he believed he was better than I was.

When I tried to talk to him about people I just met in Dungloe, he waved the conversation to a halt and told me he had left that backward place behind him and was not interested in it any more.

Later in the evening, I asked a friend about the reasons for Brendan turning himself into a Scotsman and I was told that Brendan had married a Scots girl, and as a result had a new accent, a new religion, and an Orangeman's outlook on life.

The irony was that the other Scots workers never accepted him no matter how hard he tried to ingratiate himself with them, and, of course, the Irish Catholics viewed him with utter contempt and would not even talk to him.

What I found bizarre was that Brendan had acquired that Glasgow Scots dialect that is often incomprehensible to anyone but another Scot, and in the process he had abandoned Donegal English, which at least is understood all over the British Isles.

I spoke to him several times after that and I had great difficulty keeping myself from laughing out loud at his pretensions. He talked non-stop about the Rangers football team, and he had given his sons, Angus and Billie, Scots names – trying to reinforce his Protestant/Scots identity at every turn.

Brendan was a total outsider, but did not seem to be aware of it.

He acted as if he had been a Scotsman all his life, and he obviously did not care what anyone thought of his transformation.

I think marriage had made him a little crazy.

*

Jonathan Smith, a Canadian, was as much an outsider as the nurse, but for an entirely different reason.

Smith, who was in his mid thirties, was an Anglophile who had cut all ties with his homeland because, according to him, there were too many Italians, Spanish and other nationalities being allowed into Canada. He wanted to live in the country where only British people lived.

Smith, who was of Scots and English descent, had no skills of any kind. When he arrived in London he worked as a labourer on the construction sites. After learning about the big money being made in the Highlands he had eventually found his way up there, bringing with him a pride in all things British, and a hostility to all of the 'lesser tribes'. If Smith had been living in the Nineties instead of the Fifties he would have been called a skinhead or a Neo-Nazi.

Smith would have found a lot of kindred souls in London in the Fifties, who agreed with his super-Brit sentiments and his hostility to the foreign refugees, who were also pouring into London from all over the British Empire.

Smith viewed the situation in London as much worse than the situation he had left behind him in Toronto, and for him it was a case of out of the frying pan into the fire.

He got himself into trouble right away up in the Highlands because he did not seem to realise that while the words English and British were interchangeable in London, north of the border most Scotsmen were Scots first and British second.

There was also an undercurrent of Scots hostility towards the English, which was based on the idea that the English were always giving the Scots the short end of the stick within the United Kingdom.

So, when Smith was going on about the virtues of his English cousins, he was viewed as a half-witted eccentric by his Scots listeners, and as a figure of fun by the Irish and the other nationalities in the camp.

Somebody told Smith I was planning to emigrate to the United States. When he heard it he came to me on the job one day and spent an hour trying to talk me out of it. He said America was full of Africans, Chinese, Indians and other inferior people, and the place should be avoided at all costs.

I did not respond to him because he would be the last person that I would take advice from.

I do not know whether Smith was just plain stupid or was so into denial that he could not see that the Scots did not want to have anything to do with him, never mind make a common cause with him. In truth he was an embarrassment to them, and they were extremely hostile to him, but Smith soldiered on anyway.

On Saturday nights when he went down to the canteen to have a few drinks, the Irish workers liked to encourage him to come out with his super-Brit sentiments so that the Scots who were there would get embarrassed.

The Irish would even go as far as to boast about Irish achievements just to get him involved in an argument.

In the end, however, Smith was quietly removed from the scene when he came into work one Monday morning with a hangover and was considered unfit for work. He resigned before anyone could fire him.

Smith blamed the Irish for the loss of his job; he said they resented his British patriotism. But he was pointing his finger at the wrong ethnic group. His patriotism did not bother the Irish at all – it was the Scots who could not put up with him. The Irish just loved him because he gave super-Brits a bad name.

*

The ultimate outsiders in Dalcroy were not foreigners, but a farming family who lived on a hill above Bohesbic. They seemed to have little contact with anyone in the area.

I discovered their existence because of my alienation from the food served in the mess hall.

By the time I was two years up in the Highlands, I had developed a craving for the type of food people took for granted over in Ireland.

The kind of food I longed for was fresh eggs, home made bread, fresh milk, bacon, butter, and all the other ordinary foods that I never tasted in the camp.

I had got over my craving for beef because one taste of that Argentine beef had cured me.

But I still longed for all the other food I liked so much.

Nothing was fresh in the mess hall at Dalcroy. All the milk that was served with our tea was condensed milk that came out of a can; all alleged eggs, if they were eggs and not some kind of a substitute, were scrambled into a tasteless lump; all bread was a pale mass-produced product that had no taste either; and butter and bacon were alien to the camp menu. You could fill your stomach with the camp menu, but you remained hungry for fresh food.

There were no cafés that I knew of in the Highland area where we could go to satisfy those cravings, and none of us felt like going into one of the fancy hotels in Pitlochry and spending a fortune on a meal that would satisfy our fantasies. So we did without and continued to dream.

But one evening, while I was relaxing on my bed, a delicious, mouth-watering aroma came drifting my way, which was without question the smell of frying bacon.

When I went down the hut to identify the source, I found a Scotsman frying up a meal of bacon, sausages and fresh eggs in a pan. Beside him on the bed were a big loaf of homemade bread and a chunk of butter. The sight was unbelievable.

When I asked him where he had got all this beautiful food, he said he had bought it from the Scots farmers who lived a mile up the mountain from the construction site. He said they raised chickens and pigs, grew their own wheat, and sold bacon, eggs, sausages, butter and bread to a few privileged customers.

'But they wont sell any to you,' he said. 'They do not like Irish people and they are not that fond of Lowland Scots people either, because they think the Lowland Scots look down their noses at Highlanders. They do not want Lowland Scots or Irish coming up to the farm.

'You have to get one of the teuchters to go up there for you and that is the only way you are going to get this food.'

I knew who the farmers were right away. There were two old men and a woman who used to come to a hill above the construction site and stare down at the workers for hours but never got any closer. And if anyone tried to go up and talk to them they retreated.

I also knew all about the negative view the Lowland Scots had of the islanders from the Hebrides and of those from the Highlands.

This was an animosity that went back nearly a thousand years and was similar to the antipathy the Lowland Scots had for the Irish.

I had heard that these farmers were all from the same family and were considered a bit eccentric. Nobody had mentioned the delicious food that they had up there on the mountain, and that it was being sold to select customers.

I tried my best to get one of the Teuchters in the camp to go up to the farm for me, but they were either not interested or they wanted too high a fee to run the errand. I did not know MacDonald at the time.

One day at lunchtime when there was nothing to do at the excavator, I decided to walk up the hill and see if I could buy some groceries. One of the farmers was in the backyard when I arrived, and had not seen me coming.

When he saw me and heard I wanted to buy eggs, bacon, butter and bread, he reacted with a great deal of nervousness and hurried into the house and closed the door.

I waited for twenty minutes and was at the point of walking away when he came out again with a dozen eggs, about a pound of bacon, two loafs of bread, and a big chunk of butter. He said the price was two pounds, which I gladly paid.

I asked him if I could come back again but he would not give me an answer on that.

He did tell me that he did not like too many visitors. I told him the men down on the site were very thrifty and few would come here and spend the money, so he should not worry about being overrun with visitors. He said nothing.

I really had the most memorable meal of my life that night and shared some of the delicacies with a few of my friends.

After that I would go back up there about once a month. Although I never got a warm welcome, I was always able to make a purchase.

In the months that followed, a few of my friends also went up the mountain from time to time to buy eggs, bacon and bread, and they all got the same treatment.

I often wondered what these people were really like and if they ever had a normal relationship with anyone outside the family.

And I wondered what made them so afraid of the Irish, since the Irish I worked with wouldn't harm a fly.

I never did get an answer to either question.

*

The last person on my list of outsiders was a Glasgow youth living in the next hut who had a weak mind and an obsession with fresh fish. He would go to any lengths to get it.

I do not know what his real name was but everyone called him Jock up to his face and Mad Jock behind his back.

Jock was a Catholic and his parents were from a Donegal island, on which he spent every summer when he was growing up. It was on the Donegal island that Mad Jock acquired his taste for fish, which was the sole source of protein on the island.

Jock liked to live with Donegal people because he said the other Scots were abusive when they heard about his religion and his ethnic background. He said he felt like an outsider among them. But the Irish were inclined to view him as a Scotsman, his ethnic background and religion notwithstanding. So he lived in a sort of no man's land with Jonathan Smith, the Anglophile.

But it was not his ethnic identity problems alone that made him an outsider; it was his suicidal activities on the motor cycle that he owned that set him apart from other workers and gave him the reputation of being a man whose mind was not fully functioning.

When Jock first came to Dalcroy, he used to travel seventy miles on his motor bike to the nearest port to buy fresh fish, which he cooked on the stove for breakfast and dinner.

But when he discovered that there were herring lorries that came down the main roads from Ullapool on their way to Glasgow and Edinburgh with the previous night's catch, he began to force these lorries to stop and sell him fish. He did this by getting in front of the lorries with his motorcycle and slowing down to the point that the lorry drivers had either to run over him or to stop.

I am sure that many of the lorry drivers would have liked to run over him because he was such a pest, but in the end they would stop and fill up his bucket with fish.

Although he rarely paid for the fish, Jock was not generous with them when he cooked them in the hut at night or grilled them over a fire on the job. He would say he risked his life for the fish and anyone who wanted one had to pay for it.

So he charged a modest sum for fresh herring, and you paid an additional fee if you wanted one grilled.

I also loved fish, since I also spent many long summers on Innisfree Island, where my grandparents lived, and fish was a steady diet on that island also. But I would not pay Jock for his fish and instead bartered with him: one egg for one herring. He was sure I was getting the best of the deal, but I set the terms and he could take it or leave it.

I do not know why he never bothered taking a run up the mountain to buy eggs directly. But he never did.

He once said that he didn't want to bother with those farmers because he heard they were a little strange.

When I heard that I thought that was funny because, as far as I was concerned, he was a little strange himself.

I felt sorry for Jock, the nurse, and other outsiders because as an Irishman I had a taste of what they were going through. I felt especially sorry since I had members of my own race and religion to fall back on and they had nobody at all.

They were the ultimate outsiders.

CHAPTER 21

Back to Dungloe

I TOOK MY FIRST trip back to Dungloe after spending more than two years in the Highlands. It was summer, and I was very glad to get a break from the daily grind at the construction site.

I felt mixed emotions about going home. My memories of Dungloe were not exactly happy ones. I still had bad dreams about living on the edge of destitution while keeping up the appearances of a successful business family. I remembered the bills coming in and no money to pay them. I remembered the envy I had of families with a steady income who appeared to live a 'normal' life. I remember how I envied those who had a steady income from a job, no matter how small that income.

All of this was very fresh in my memory and I was not that anxious to go back to the scene of such trauma, but I needed a break from the Highlands and I went anyway.

I took the Derry Boat back to Ireland and partied along the way with the other tunnel tigers returning home. Six of us hired a taxi in Derry and headed for Dungloe, making a stop at Jackson's Hotel, Ballybofey, for refreshments before going on the final leg of our journey to Dungloe.

When I entered the town I saw that little had changed. There were no new buildings, the same people, and the atmosphere of a community just drifting along.

I once had a very close relationship with this place. This town had been the only home I had ever known, and before I went to Scotland I had been part of its fabric. But I now knew a separation had taken place. I had gone away from home in more senses than one.

I now saw the town with detachment, casting a critical eye on the older buildings, including my own hotel. The hotel for some reason seemed smaller than I had remembered it, as did many of the buildings on Main Street.

I saw a familiar sight as we arrived in town: a group of men were standing talking to one another at Mulhearn's Corner. These were men who hung out there every day: they had been there when I left; they were still there.

It seemed very odd to me to be a customer in my own bar. The first night I went into the bar, there was a young bartender whom I had never met before. He did not know me and treated me as if I was a stranger.

I went around to many of the other eleven bars in the town during the next few days, many of which I had never been in before. I had plenty of money and I enjoyed the experience. After a while the bar hopping depressed me. I realised that the owners of these bars had been able to hang in and continue to make a living in Dungloe, whereas I had had to go abroad to make ends meet. I had the feeling of being a loser, and that everyone knew I was a loser.

In my heart I knew the failure of the business had nothing to do with me, or my parents, because they had grown old and were unable to run it. But it hurt me to see our tenants doing so well in the bar – it appeared to be full every night that I was there – and they were obviously making a big success where I had failed.

One evening I bought a bottle of whiskey and went down to see Andy Campbell, who was also home on holiday. He lived in a big, beautifully furnished two-storey house outside Dungloe. He was delighted to see me and was the same friendly Andy Campbell I remembered from my days in the bar.

There was none of the cool impersonal attitude I had witnessed up in Dalcroy and we talked casually about a whole variety of subjects, none of them related to work.

We chatted for a while and then he opened the bottle of whiskey and asked me if I wanted some. I said no – that I preferred Guinness.

Andy poured himself a huge tumbler of whiskey and then drank it down as if it were water. Then he filled the glass up again and put the bottle away.

'Have you seen your Uncle Owen yet?' he asked. 'I saw him yesterday and he said that you had not been to see him.'

I told him that I had been down one day but that he was not around.

'Be sure you visit him before you leave Dungloe.'

When I was leaving he told me that I should not continue to work in the Highlands.

'You are wasting your time up there, no matter how much money you are making. There is no future there for you. Go to America. There is more opportunity there. You could even go part-time to university. A cousin of my wife's did that and he became a teacher. My son Michael is going out. I told him to go and not to be hanging around the Black Gang getting into trouble.'

I agreed with him and said I had been thinking about it for some time. I told him that I wanted to go to university and it was just not possible in Scotland or Ireland.

Only in America could I work during the day and go to university in the evenings.

*

Guilt forced me to return to my Uncle Owen's house that evening. He was at home. I had never really been very close to Owen and I couldn't remember ever having a heart to heart conversation with him on any subject. He had an opinion on every issue and he refused to accept any opinion that was in conflict with his own. He used to annoy my father because he was constantly arguing with him.

Owen lived in a ramshackle thatched cottage in which he and my father had been born. The place was dark and depressing, and Owen would never have won any prizes as a housekeeper.

Owen was seventy and had worked in Scotland in his early years and later on emigrated to the United States, where he had worked for many years doing odd jobs. He had never saved any money in his younger days, and he had not saved any money when he was living in the United States, and in the end my father, who was working in the United States and who saved every penny, sent Owen back to Donegal to live in the cottage where he was born.

Part of Owen's problem was that he was a little too fond of liquor. He was a binge drinker who went on a binge every six months and was a teetotaler in between. My father viewed Owen as a cross he had to bear and was disgusted at his drinking habits.

Owen was cold sober when I arrived and we had a great conversation for several hours.

Owen had read widely over the years and behind that rough exterior was a sharp, intelligent mind. He could discuss anything, from politics to literature.

If Owen had been born in the 1970s instead of in the 1880s he could have availed himself of scholarships and God knows how far he would have gone. He certainly had the brains and the interest in the world around him.

But he never had the opportunity to improve his lot and he ended his life as he began it – penniless.

I knew better than to bring a bottle of whiskey to Owen, but on the other hand I did not want to leave him without giving him something, because I knew what it was like to be poor.

So I handed Owen ten pounds, which was a considerable sum of money in those days. I resisted the temptation to tell him not to use the money to finance a drinking binge.

After all, I knew that he had probably heard that I had been pub crawling in Dungloe during the last several days, so I was in no position to talk about the evils of booze. It would be like the pot calling the kettle black.

Owen was delighted with the gift and got very emotional. It was the first time that I had ever given him a gift, and he kept thanking me as I left.

I was a little embarrassed that this new bond with Owen had been established with money, but I did not care because I was glad I had made him happy. The man to man conversation had also made me happy because it was the first time I had had such a conversation with Owen.

The visit to Owen had given me some food for thought. Here was a bright, intelligent man who had never made anything of his life because he just drifted from one job to another, like one of those long distance kiddies.

Prior to my visit with Owen I had been tempted to spend a considerable amount of my bankroll in the bars before I returned to the Highlands, but I slowed down and went back to Scotland with plenty of money to spare.

I even had enough money left to send twenty pounds to my parents in Galway, and what was left I used to open a bank account in Scotland.

I met Owen several times before I went back and he was a model of good behaviour.

Owen did not go on a binge while I was in Dungloe, which was a relief to me, but I heard afterwards that the day I went on the Derry Boat he was making the rounds of the bars, drinking up a storm and boasting about his nephew who was making a fortune in the tunnels in Scotland.

Owen had a habit of exaggerating. By the end of his binge he had me running one of the construction sites in the Highlands.

Somehow I did not feel guilty when I heard about Owen's drinking. Owen had a terrible life and a two-week binge was not going to make it any worse.

On these binges, he would drink until all the money was gone, and then he would stop.

CHAPTER 22

Glasgow

ON MY WAY BACK from Dungloe I decided to spend a few days in Glasgow visiting two relatives who had lived there for years. It was a decision I made on impulse, so the relatives were not expecting me.

I had been close to these cousins and I wanted to see them before I headed off to the United States. I knew that once I emigrated to the United States the chances were slim that I would come to Glasgow again.

I had their addresses and I thought that I would have no trouble finding them. They lived in an area known as the Gorbals; a poor, tough neighbourhood where the majority of the people were of Irish extraction.

I had two places to visit: one a boarding house run by a cousin; the second an apartment where another cousin lived.

I decided to visit the boarding house first because I thought that my cousin would have plenty of room to keep me.

Somehow or other I had visualised the boarding house as something like Campbell's Hotel back in Dungloe. But when I arrived at the street, which was comprised entirely of tenement buildings, the street number was on a massive rickety-looking building that had no entrance on the main street.

There was an alleyway on one side of the building and there I found a door that opened into the side of the building. I opened it and found myself in a corridor that had a number of doors along one side of the wall. Obviously I was in a building that contained multiple living units on multiple floors.

The whole building had that abandoned look about it that is the result of being badly maintained.

It took me twenty minutes to find my cousin's apartment, which was on the fifth floor. I discovered it contained four seedy rooms and a small kitchen. There were three beds in each room.

My cousin was delighted to see me. When I told her that I was just passing through on my way to the Highlands she invited me to stay as long as I wanted.

I was amazed at her appearance. Although she was nine years older than me, when I was in my early teens I had a big crush on her and I thought she was a knockout. Now, however, her face was lined and her hair was going grey, and she obviously had no interest in her personal appearance.

She told me she had fourteen boarders who were labourers in the Glasgow area. She fed them breakfast, evening tea and a meal that they carried to the job for midday.

'My Scotch husband is too fond of the bottle, and only for the boarders I would starve.'

Later on that evening I met the husband, who was a bad-humoured Scotsman who did not utter a word to anyone all night. He just sat by the fire nibbling out of a bottle of some type of liquor. There was obviously no romance left in this marriage because the pair of them did not even look at one another all evening.

I found the sleeping arrangements a little unusual, to say the least. My cousin and her husband slept in one room, and the boarders slept two to a bed in the other three rooms. A single bed was set up in the kitchen for me, and the boarders teased me all evening about the fact that I was a guest and had my own room.

The meals put on the table that evening were excellent. A steak and kidney pie was served, with plenty of potatoes and bread served on the side. Then there was a pudding, and plenty of warm, well-brewed tea.

The conversation was very good too and I soon learned all the men present were eager to learn all about work in the Highlands, although few of them showed any interest in going up there.

But there was a smell of gas in the apartment that came from the gas stove, and this odour was blended with the odour of dirty feet and the smell of food, and I did not sleep a wink all night, even though I had been out in a rowdy bar for several hours and got half stewed while watching several brawls.

I wondered how my cousin had got herself in such a predicament. I knew that she had left Ireland for the same reason that I had: she wanted to find a better standard of living. But her lifestyle in Glasgow was about as low as a lifestyle could get, and no matter what her predicament had been in Ireland it certainly would not have been worse than this.

I wanted to give her money the following morning but she would have none of it. I was her guest and that was that.

*

My other cousin lived several streets away in the same area and I decided to visit her next before I headed off to the Highlands. I walked through mean crowded streets full of rowdy, ragged children on my way to my cousin's apartment. None of them paid any attention to me as I strolled along carrying my bag, and the adults on the streets ignored me also. I suppose they were all used to the sight of 'Paddys' blowing into town.

After my experience the previous night, I expected to be greeted by another dilapidated apartment that reflected the worst type of urban poverty, but I was only half-right.

This cousin also lived in one of those Gorbals tenements and it too was very crowded with occupants, but the similarity ended there, because this was a private home that housed eight children and their parents, and it was a home that was scrupulously clean. The children, while they were rambunctious, were well groomed and healthy looking, and there were no boarders in this house, so obviously my cousin had a steady income from another source.

However, I discovered later that the home was not without problems. The Scots-born husband drank much of what he earned, but the shortfall was made up by the older children, who had part time jobs and gave what they earned to their mother. The mother in turn was extremely thrifty, buying the clothes for the children at rummage sales and accepting free clothes from St Vincent de Paul.

My cousin told me about her husband's drinking problem, but

she said she did not fight with him about it because the children loved him. She thought it better to have a quarter husband than no husband at all.

It was a tough existence. Every penny that came into the house was spent to survive. There was not a penny left for toys or chocolate, and hunger was only kept at bay by the pennies earned by the children.

That afternoon I went out to one of the stores and bought scores of chocolate bars and other delicacies to distribute among the children. I was delighted by how much they enjoyed the treat, but I do not think their mother was that happy with my generosity – probably because the children might want these treats on a regular basis once they had tasted them – but she made no attempt to take the sweets away from the children.

We had a long talk that night about Dungloe and the life she had left behind her. She was nostalgic when she talked about her early days in Dungloe, and it was obvious to me why she felt that way. At one point she asked a question that really took me by surprise. She asked if I thought I would ever make it to ganger (foreman) on the job in the Highlands. I really did not know how to answer her, because being a foreman up in the tunnels was certainly not something I was aspiring to.

I thought of getting a degree from a university in the United States, and I thought of working in the business world once I got that degree. But whatever I did in the future I certainly hoped that the job would be a little more interesting than a ganger job.

However her question was an indication of how people were viewing me in light of my new position as tunnel tiger. Before I went to Scotland people assumed I was going to university and would wind up in one of the professions. Now, given the fact I was working as a labourer, it was assumed that the best I could hope for was promotion to a ganger

I had a good rest that night sleeping with the boys in their room, but in a perverse sort of way I was looking forward to returning to my single bed in the Highlands, which I had all to myself. It seemed more private, even though there were forty other beds in the hut.

*

My brief visit to Ireland and the stopover in Glasgow had given me food for thought. I realised that I could not afford to drift along from year to year if I wanted to make something of my life. I had to make some plans for the future. I knew I had to come to some decision about what my next move was going to be. I could see that Uncle Owen's plight was a direct result of leading a life that had no focus to it. The only planning that Owen was ever involved in was how to get enough money to finance the next binge.

My cousins in Glasgow had wound up at the bottom of the heap also, and alcohol was involved there too.

But their predicament involved more than alcohol abuse: neither Owen nor the two husbands in Glasgow had any ambition or an education or skills of any kind, and their lives had been a journey from one menial job to the next without ever a hope of getting ahead in the game, not even for a short period of time. They had been born at the bottom of the heap, and they believed that existing at the bottom of the heap was a natural state of affairs. I wondered if any one of them had ever made any effort to change the course of their lives by taking some initiatives.

The only reason that members of my immediate family were ambitious about acquiring a college education was the example set by my parents who knew the great value of an education. They had been determined to put every member of the family through university. They had managed to educate the three oldest girls and they became teachers – a highly respected profession in Ireland – and as a result my sisters had jobs that provided a secure and steady income.

And even though they had been unable to pay for a higher education for the younger members of the family, my parents still encouraged us to get an education any way we could.

*

Most of those I met in Scotland who had skills were able to escape poverty of the type I had seen in Glasgow and were living in comfortable homes outside Glasgow. All those with an education had long left the Gorbals behind them.

As I headed back to the Highlands on the train the following morning I made up my mind that I would never wind up like Owen or my cousins in the Scottish Lowlands. I had seen what the end of that road looked like, and I did not like what I had seen.

As the train rolled northwards towards Pitlochry the following morning, I decided that I would spend no more than one year at Dalcroy, and then I would head off towards the United States.

I also decided that I would limit my recreational trips to Perth and Aberfeldy at the weekends, and stay out of the canteen as much as possible. If I did this I would accumulate money at a greater rate than if I spent a considerable chunk of my resources on entertainment.

Once I had a plan in place, I felt much more relaxed about going back to the Highlands. I knew I had been just drifting along up to that point and that it might just have taken years to get to America had I not focused on a given plan of action.

CHAPTER 23

End of an Era

I SETTLED DOWN TO work and to save during my last twelve months in the Highlands and I never felt more at ease in my life. I also explored other aspects of the cultural life in Scotland, especially the pipe band competitions and the step dancing competitions, which were very enjoyable. And I explored the countryside around Perthshire and the many historic castles in the area.

More importantly I learned to ignore the negative aspects of life in Scotland, like the snide remarks made about my race and my religion, and towards the end of my stay in Scotland they never angered me like they had in the beginning.

Perhaps I was growing older and more mature and I was able to trade off the positive aspects of my stay in the Highlands against the negative, and I could come to the conclusion that overall life in Scotland was not that bad after all.

Deep down I knew I was grateful for the money I had earned and the freedom this money gave me to move on to the places in the world that I was eager to visit. And I was proud to have been part of those historic hydroelectric projects, which play such an important role in the Scottish economy.

As I prepared to leave for the United States, I exchanged correspondence with relatives out there in order to determine what I might expect as an unskilled, uneducated person, and I was told Americans as a whole were friendly to the Irish, but as far as jobs were concerned, I would have to start at the bottom of the heap – maybe with a pick and shovel. There was no Andy Campbell there to help me in the State of New Jersey.

I didn't mind the pick and shovel work as long as there was opportunity to move upwards. Obviously there was, since both my older brothers had already obtained college degrees at nights while working at menial jobs during the day.

The changeover from life in the Highlands to life in the New York Metropolitan area came very abruptly. And dramatically.

I worked at Bohesbic up to the very last day in Scotland and then handed in my resignation and collected my final wage packet.

The following morning, at 6am, I was on a bus that was headed for Glasgow. At 1pm I boarded a plane for New York's Idyllwild Airport (JFK). I arrived in New York sixteen hours later, after a stop over in Gander, Newfoundland

My brother Bernie was waiting for me at the airport and he escorted me out to his car, which was a huge convertible, and soon I was travelling in towards Manhattan on the Long Island Expressway, which was choked with traffic. I had never seen so many cars on one highway in all my life.

The sights and sounds of Manhattan were overwhelming. The noise, the throngs of people, the bright lights of Broadway were so different from either Ireland or the Highlands. It seemed like I was on another planet.

As we headed through the Holland Tunnel towards Bayonne, New Jersey, where my sister Rose and her husband Jimmie were waiting for me, I remembered what John Duffy had said to me on my first day at Dalcroy:

'Things might seem strange now, but in a week or so you will get used to it.'

I did.

Bibliography

Books about North West Donegal

Campbell, Patrick, *Death in Templecrone*
The Last Days of Oscar Devenney
Memories of Dungloe
A Molly Maguire Story
Devenney, Donnchadh, *Footprints Through the Rosses*
Doherty, Jenni and Liz, *That Land Beyond: Folklore of Donegal*
Dolan, Liam, *The Third Earl of Leitrim*
Gallagher, Barney, *Arranmore Links*
Gallagher, Francis, *Leaving Ireland*
Gallagher, Patrick, *Paddy the Cope*
Hill, Lord George, *Facts From Gweedore*
Holland, Dennis, *The Landlords in Donegal*
MacGill, Patrick, *Moleskin Joe*
The Great Push
Children of the Dead End
McMahon, Sean, *Light on Illancrone*
O Gallchóbhair, Proinnsias, *The History of Landlordism in Donegal*
Nic Giolla Easpaig, Áine and Eibhlin, *Sisters in Cells*
O'Donnell, Peadar, *The Knife*

Books on the Pitlochry Area

North of Scotland Hydro-Electric Board: *Power from the Glens,* 1973
North of Scotland Hydro-Electric Board: *The Tummel-Garry Hydro-Electric Schemes,* 1950
North of Scotland Hydro-Electric Board: *Tummel Valley Hydro-Electric Schemes,* 1950
The Pitlochry and District Tourist Association: *The History and Heritage of Pitlochry and District,* 1974
Sinclair, Duncan McDonald, *By Tummel and Loch Rannoch*. Perth: Northern Print Services, 1989

Index